Oxford Chemistry Serie

General Editors

P. W. ATKINS J. S. E. HOLKER A. K. HOLLIDAY

R. J. PUDDEPHATT

LECTURER IN CHEMISTRY, DEPARTMENT OF INORGANIC, PHYSICAL, AND INDUSTRIAL CHEMISTRY, UNIVERSITY OF LIVERPOOL

The Periodic Table of the Elements

Clarendon Press · Oxford

Oxford University Press, Walton Street, Oxford OX2 6DP

OXFORD LONDON GLASGOW
NEW YORK TORONTO MELBOURNE WELLINGTON
KUALA LUMPUR SINGAPORE JAKARTA HONG KONG TOKYO
DELHI BOMBAY CALCUTTA MADRAS KARACHI
IBADAN NAIROBI DAR ES SALAAM CAPE TOWN

FIRST PUBLISHED 1972
REPRINTED (WITH PROBLEMS) 1972, 1978

PRINTED IN GREAT BRITAIN BY
J. W. ARROWSMITH LTD., BRISTOL, ENGLAND

Editor's Foreword

THE periodic table of the chemical elements has existed for just over a century. During that time it has provided a basis, and also a stimulus, for enormous developments in inorganic chemistry, and it continues to play an important part in the further extension of the subject. Increasingly, however, it must also provide a framework which students can use to gain some knowledge and understanding of the vast chemistry of more than one hundred elements and their compounds. This book is designed to provide, without a detailed discussion of theoretical concepts, an account of the way in which the periodic table is constructed, and of the trends in properties, both physical and chemical, which are revealed by the table.

The framework provided by the periodic table for the concepts of inorganic chemistry will be amplified and extended in books on main-group and transition-element chemistry as well as in a treatment of ligand-field theory. Further books will describe the electronic structures of molecules and of solids, and the biological and geochemical significance of inorganic systems.

A.K.H.

Acknowledgments

The author wishes to thank Professor A. K. Holliday, who suggested that this book should be written and suggested several improvements to the text. He is also grateful to Professor C. E. H. Bawn for his encouragement, and to the Clarendon Press for many helpful comments. Thanks are also due to the authors and publishers who permitted reproduction of diagrams; specific acknowledgments are given in the text. Finally, he would like to thank his wife for her assistance at various stages of the book's preparation.

Contents

1. Introduction

THE first periodic table of the elements was drawn up by the Russian chemist Mendeleev in the years 1868–1870; one version of Mendeleev's periodic table is shown in Fig. 1. It was obtained by arranging the chemical elements in order of increasing atomic weights. Vertical columns of elements were called *groups*, horizontal rows of elements, *periods*. Elements in the same vertical group possessed similar chemical properties.

This discovery of the periodic behaviour of the chemical elements provided an important stimulus for further chemical investigations. In the time of Mendeleev there were many 'holes' in the periodic table, where elements had not yet been discovered. Mendeleev was able to predict the chemical properties of each of these elements on the basis of the position in the periodic table. His predictions were found to be amazingly accurate when the elements were later discovered and their chemical properties investigated.

However it soon became apparent that a periodic table based on the atomic weights of the elements was not entirely satisfactory. For instance it was shown that tellurium had a higher atomic weight than iodine despite the fact that the opposite arrangement in the periodic table was clearly required by their chemical properties.

Series	Group I — R_2O	Group II — RO	Group III — R_2O_3	Group IV RH_4 RO_2	Group V RH_3 R_2O_5	Group VI RH_2 RO_3	Group VII RH R_2O_7	Group VIII — RO_4
1	H=1							
2	Li=7	Be=9	B=11	C=12	N=14	O=16	F=19	
3	Na=23	Mg=24	Al=27	Si=28	P=31	S=32	Cl=35·5	
4	K=39	Ca=40	?=44	Ti=48	V=51	Cr=52	Mn=55	Fe=56, Co=59, Ni=59
5	Cu=63	Zn=65	?=68	?=72	As=75	Se=78	Br=80	
6	Rb=85	Sr=87	Yt=88	Zr=90	Nb=94	Mo=96	?=100	Ru=104, Rh=104, Pd=106
7	Ag=108	Cd=112	In=113	Sn=118	Sb=122	Te=128	I=127	
8	Cs=133	Ba=137	Di=138	Ce=140				
9								
10			Er=178	La=180	Ta=182	W=184		Os=195, Ir=197, Pt=198
11	Au=199	Hg=200	Tl=204	Pb=207	Bi=208			
12				Th=231		U=240		

FIG. 1. Mendeleev's periodic table. The notation used indicates the atomic weights of the elements, or the predicted values for elements which were still to be discovered; for instance, ? = 68 for the element now known as gallium.

It was not until the electronic structures of atoms had been established that the full significance of the periodic table could be recognized. A brief account of our present knowledge about atomic structure is given in Chapter 2, and the way in which the modern periodic table is built up logically on the basis of this knowledge is discussed in Chapter 3.

Following the establishment of the periodic table on the basis of the electronic structures of the element atoms, rather than on the basis of their atomic weights, its importance in chemistry has steadily grown, and two aspects in particular deserve special mention.

First of all, there has been a great increase in recent years in our knowledge of the physical properties of the chemical elements and their compounds, largely due to the use of new structural methods. These properties can often be readily understood in terms of the electronic structures of the elements and their positions in the periodic table. In turn these physical properties provide a basis on which the chemical properties of the elements and their compounds can be rationalized. Periodic trends in the physical properties of the elements are discussed in Chapters 4 and 5 while subsequent chapters are concerned with periodic trends in chemical properties of the elements and their compounds.

The second important aspect of the periodic table is seen in its almost universal application in inorganic chemistry. The periodic table has played a major part in the great resurgence of interest in inorganic chemistry; the discovery of new elements and compounds has been stimulated on countless occasions by a study of the periodic table and its possible amplification and extension. Moreover, the vast expansion of knowledge about the elements and their compounds can only be comprehended in terms of the periodic table; it is no longer possible, or even desirable, to learn the properties of each chemical element or compound in isolation. Because broad trends in chemical behaviour can readily be rationalized using the periodic table, inorganic chemistry need no longer be regarded as a mere catalogue of facts. In this short book it is hoped to show, by concentrating on *comparisons* between and *trends* in the chemical properties of the elements in terms of their positions in the periodic table, rather than on individual compounds, that the study of inorganic chemistry can be a fascinating and stimulating subject.

2. Atomic Structure

THE modern periodic classification of the elements depends upon atomic structure, and in particular upon the way in which the energy levels of electrons in atoms are arranged. These energy levels are defined in terms of whole numbers, called *quantum numbers*, and some understanding of their origin is necessary before the detailed structure of the periodic table is discussed further.

The hydrogen atom with one electron is the simplest possible atom and has served as a model for all modern theories of atomic structure. Niels Bohr in 1913 put forward the first theory of the hydrogen atom, making use of the quantum theory put forward by Planck a few years before.

Bohr theory of the hydrogen atom

The Bohr model of the hydrogen atom may be pictured as the electron moving in a circular orbit about the proton like the earth circling the sun (Fig. 2).

The electrostatic attractive force between the electron and the proton is balanced by the centrifugal force on the electron due to its circular motion. This leads naturally to the equation

$$\frac{1}{4\pi\varepsilon_0} \cdot \frac{e^2}{r^2} = \frac{m_e v^2}{r} \tag{1}$$

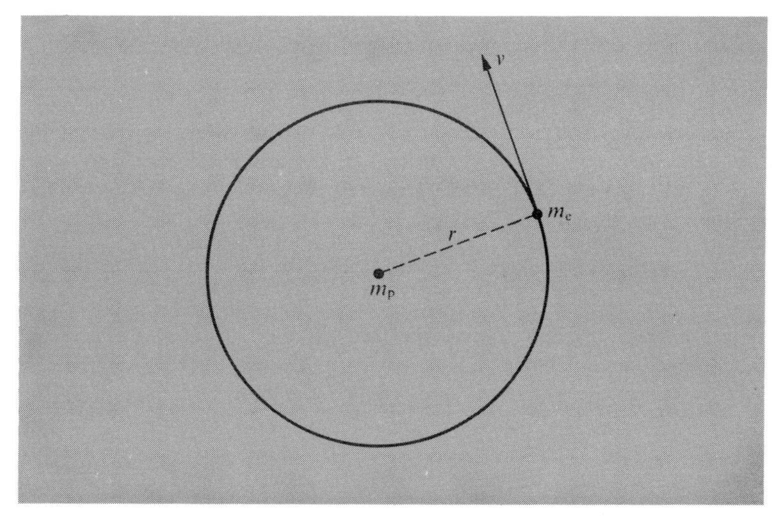

FIG. 2. The Bohr model of the hydrogen atom.

where e, m_e and v are respectively the charge, mass, and velocity of the electron and r is the radius of the orbit as defined in Fig. 2. ε_0 is a constant known as the permittivity of free space.

So far the picture is the same as one proposed earlier by Rutherford in which the orbital radius r could take any value. Bohr now made the apparently startling assumption that the angular momentum of the electron, $m_e vr$, was quantized, that is that it could take only certain discrete values. So

$$m_e vr = n \cdot \frac{h}{2\pi},$$

where h is a fundamental constant called Planck's constant and n is an integer ($n = 1, 2, 3, \ldots$). Substituting for v in equation 1 we obtain

$$\frac{1}{4\pi\varepsilon_0} \cdot \frac{e^2}{r^2} = \frac{m_e}{r} \cdot \frac{n^2h^2}{4\pi^2 m_e^2 r^2},$$

which on rearrangement gives

$$r = \frac{n^2h^2\varepsilon_0}{\pi m_e e^2}.$$

This equation then gives the allowed radii of the electron orbits in the hydrogen atom in terms of the *quantum number*, n. For each allowed radius r_n there is an allowed energy of the electron E_n, i.e. $E_1, E_2, \ldots$, which can easily be calculated.

Bohr's theory was most successful in rationalizing known facts concerning the hydrogen atom. However the theory was not so successful when applied to atoms with more than one electron and it was soon modified and then discarded.

One of the chief objections to the Bohr theory was the rather arbitrary way in which quantum numbers were introduced. Another difficulty was that the theory assumed the electron to behave only as a particle. Future theories were to take account of the wave properties of the electron from which the quantum numbers evolve naturally.

The wave theory of electrons

At the time when Bohr proposed his theory of the hydrogen atom it was already known that the photon, the smallest unit of light, showed both particle and wave properties. The energy of the photon is given by the Planck–Einstein expressions,

$$E = h\nu = mc^2,$$

where h is Planck's constant, ν the frequency of the light wave, m the mass of the photon, and c the velocity of light.

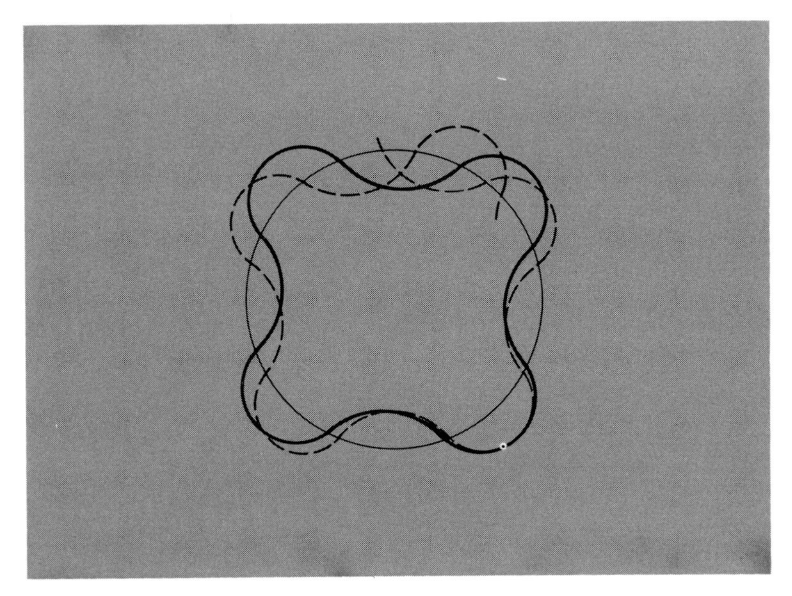

FIG. 3. Representation of an electron wave. The solid line represents a possible stationary wave, while the dotted line shows how a wave of somewhat different wavelength would be destroyed by interference.

Now the momentum p of the photon is given by

$$p = mc = \frac{hv}{c}.$$

Since the wavelength λ and frequency v of light are related by the expression $\lambda = c/v$, we have

$$\lambda = \frac{c}{v} = \frac{h}{p}.$$

De Broglie suggested that this expression, which gives the wavelength associated with a particle having momentum p, should hold for all matter, and in particular that it should hold for an electron in a 'Bohr orbit'. In such a situation it is necessary to have a standing wave, i.e. there must be an integral number of waves in the Bohr orbit (since otherwise the wave would be destroyed by interference, as shown in Fig. 3). Thus we get the expression

$$2\pi r = n\lambda = n \,.\, \frac{h}{p},$$

which on rearrangement gives the following expression for the angular momentum rp of the electron in a Bohr orbit:

$$rp = n \, . \frac{h}{2\pi}.$$

Thus the quantization of angular momentum which was assumed arbitrarily by Bohr follows naturally from this wave treatment. It also follows that the electronic energy levels must be quantized.

The Schrödinger equation

Since an electron has wave properties, its behaviour can be described in terms of a wave function ψ. The physical significance of ψ is not easily grasped but it is best thought of as analogous to the amplitude of the wave. The probability of finding an electron in any volume is proportional to the integral of the square of the absolute value of the wave function over that volume, or probability $(x, y, z) \propto |\psi(x, y, z)|^2$.

The relationship between the energy of the electron in the hydrogen atom and its wave function was first given by Schrödinger. This relationship forms the basis of modern quantum mechanics and is known as the Schrödinger equation.

$$\mathrm{H}\psi = E\psi$$

where H is a mathematical operator called the Hamiltonian operator and represents the general form of the kinetic and potential energies of the system and E is the numerical value of the energy for a given wave function ψ.

The Schrödinger equation can be solved for certain wave functions ψ and these are known as *eigenfunctions*. The corresponding allowed energy levels are called *eigenvalues*.

$$\psi_{n,l,m}(r, \theta, \phi) = R_{n,l}(r)\Phi_{l,m}(\theta, \phi).$$

If written in terms of polar co-ordinates r, θ, ϕ the wave equation can be split up into a radial component $R_{n,l}(r)$ and an angular component $\Phi_{l,m}(\theta, \phi)$. The square of the radial part of the wave function $|R_{n,l}(r)|^2$ gives the probability of finding the electron at a distance r from the nucleus. The wave function can only be an eigenfunction when n is an integer; n is called the principal quantum number and defines the mean separation of the electron from the nucleus. Similarly the wave function can only be an eigenfunction when l is an integer. l is known as the azimuthal quantum number and defines the angular momentum of the electron; it can take the values $l = 0, 1, 2$, up to $(n-1)$.

The angular part of the wave function defines a third quantum number m called the magnetic quantum number. This gives the allowed values of the component of the electronic angular momentum in a given direction when a magnetic field is applied and can take all integral values between l and $-l$.

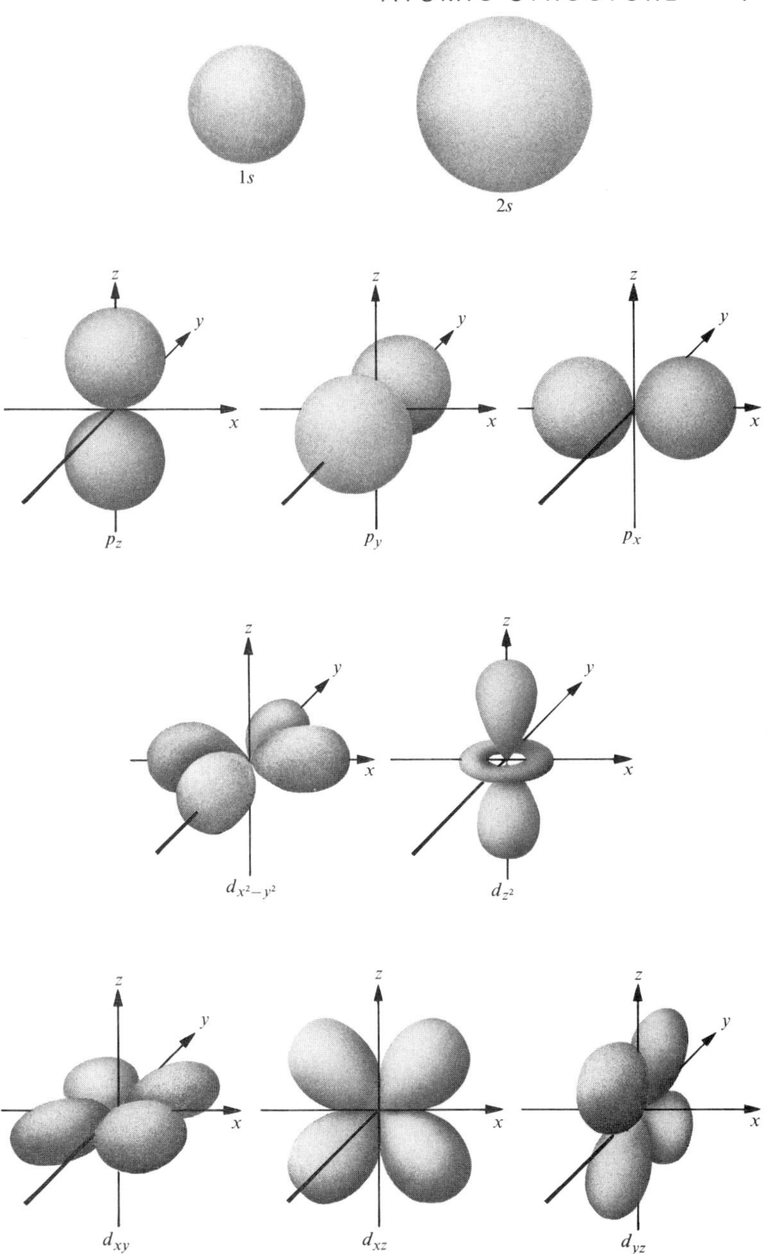

FIG. 4. Representations of atomic orbitals.

Atomic orbitals

The hydrogen wave functions $\psi_{n,l,m}$ are known as *orbitals*. The orbitals are broadly classified according to the value of l. When $l = 0$ we have an s orbital which is spherically symmetrical, when $l = 1$ we have a p orbital, when $l = 2$ a d orbital, and when $l = 3$ an f orbital. The orbitals are then prefixed by the value of n, so that for instance if $n = 2$, $l = 1$ we have a $2p$ orbital.

The p orbital can have various orientations in space depending on the value of m which can be $1, 0$, or -1. Similarly there are *five* possible orientations of d orbitals, as $m = 2, 1, 0, -1$, or -2, and *seven* possible orientations of f orbitals.

The orbitals may be represented by drawings showing the regions of space in which there is a high probability of finding the electron†. Examples of these representations of s, p, and d orbitals are given in Fig. 4. They show very clearly the directional character of the p and d orbitals.

The spin quantum number

Just as the earth spins about its own axis while circling the sun, so may the electron be regarded as spinning about its own axis. The angular momentum due to this spin is quantized so that its component in the direction of a magnetic field is given by $s \,.\, h/2\pi$, where s is the spin quantum number and can take only the values $+\frac{1}{2}$ or $-\frac{1}{2}$.

Overall then the state of the electron in a hydrogen atom is defined by *four* quantum numbers. Three of these, n, l, and m, are associated with the motion of the electron about the proton while the fourth, s, is associated with the spin of the electron about its own axis.

PROBLEM

2.1. Write down the possible quantum numbers n, l, m, and s for all (a) $2s$ (b) $2p$ (c) $3d$ orbitals. If no two electrons can have all four quantum numbers the same, what is the maximum number of electrons that can be accommodated in these orbitals?

† These drawings represent only the directional character of the orbitals and are not intended to represent the actual shapes of the orbitals.

3. Building up the Periodic Table

So far we have shown that the energy levels of an electron in a hydrogen atom are defined by four quantum numbers n, l, m, and s. The electron will normally be in its lowest energy or *ground state*; this state is the state with $n = 1$ and $l = 0$ so that in the ground state the electron will be in the $1s$ orbital.

For more complex atoms we can still define the electronic energy levels in terms of the four quantum numbers, and we shall therefore want to know the values of these quantum numbers for all the electrons in the atom. There is an important principle first defined by Pauli which determines the permitted quantum numbers for the electrons in an atom.

The Pauli Exclusion Principle

The Exclusion Principle states that no two electrons in the same atom can have all four quantum numbers the same. Thus in the helium atom, which has two protons and two electrons, both electrons can enter the $1s$ orbital; one will have the quantum numbers $n = 1, l = 0, m = 0, s = +\frac{1}{2}$, and the other $n = 1, l = 0, m = 0, s = -\frac{1}{2}$. According to the exclusion principle the $1s$ *electron shell* is now filled. In the lithium atom, which has three electrons, the first two electrons fill the $1s$ orbital so that the third electron must be accommodated in an orbital with the principle quantum number $n = 2$. When $n = 2$, l can take the values 0 or 1 so that the next electron could enter either the $2s$ or the $2p$ orbital.

Stability of atomic orbitals

We have already seen that the $1s$ orbital is more stable than the $2s$ orbital because the electron in the $1s$ orbital is on average closer to the nucleus and so feels a greater electrostatic attraction to the nucleus than the $2s$ electron. For the hydrogen atom the $2s$ and $2p$ orbitals have identical energies but this is not the case for larger atoms.

Let us look at the lithium atom in more detail. The two $1s$ electrons are directly attracted by the nuclear charge of the three protons and are very firmly held. An electron in the outer orbital with $n = 2$ has its greatest density outside the $1s$ shell and so is *shielded* from the nucleus by the two $1s$ electrons. It therefore 'feels' an *effective nuclear charge* of only a single proton and is much more weakly held. This is a somewhat oversimplified picture; the $n = 2$ electron cloud *penetrates* to some extent within the boundary of the $1s$ orbital, and this component is naturally more strongly held to the nucleus.

Now the $2s$ orbital penetrates the $1s$ to a greater extent than does the $2p$ (this can be seen in a qualitative way by inspection of the representations of orbitals in Fig. 4, p. 7) and so the $2s$ orbital is more stable than the $2p$ in the

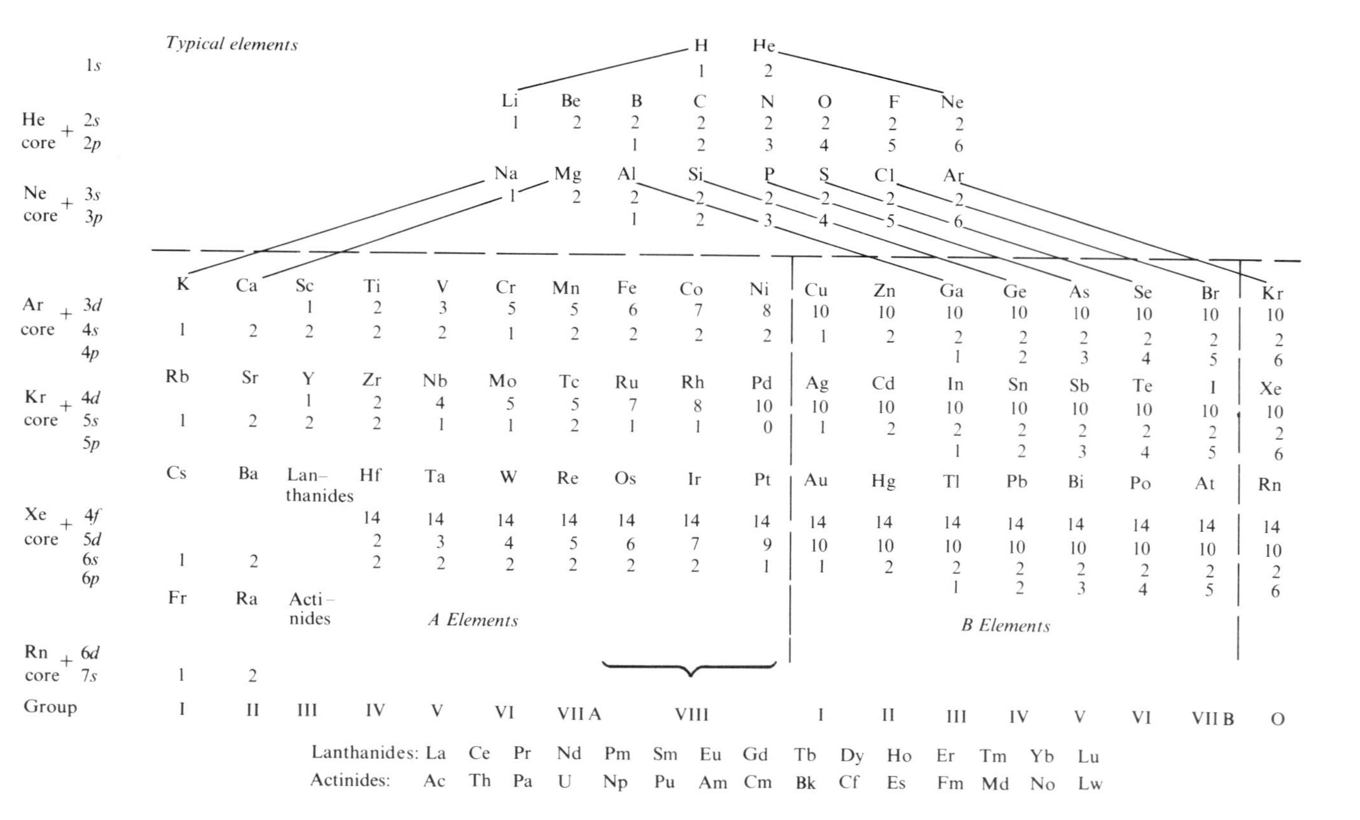

FIG. 5. A modern periodic table of the elements (after Phillips and Williams, 1965).

lithium atom. Thus the *electron configuration*, that is the way in which the electrons are distributed among the atomic orbitals, for the lithium atom in its ground state is $1s^2 2s^1$, the occupation numbers of the orbitals being written as superscripts.

In general the stability of orbitals in multi-electron atoms can be summarized as follows: if l is constant the orbital with lower principal quantum number is more stable, so we have the stability series for example $1s > 2s > 3s > 4s$. Similarly if n is constant the orbital with lower l is more stable, so the stability series $4s > 4p > 4d > 4f$ arises.

The periodic table

We have now established the principle by which the periodic table is built up. Each new element in the table contains one more nuclear charge than the preceding element; this charge is neutralized by addition of an electron which is fed into the lowest energy orbital available, taking account of the Pauli Exclusion Principle. The elements can then *be classified naturally into periods depending on which electron shell is being filled.* Each period begins with the occupation of an ns orbital and ends when the np orbitals are full.

A modern version of the periodic table is shown in Fig. 5. In the following sections we shall discuss its structure in some detail.

The short periods

The first period contains only the elements hydrogen and helium. Of all the element atoms helium has the least ability to take part in chemical combination, having very little tendency either to gain or to lose an electron (see ionization energies in Fig. 7, p. 16). For this reason helium is known as a *noble* (chemically unreactive) gas. We shall see that all subsequent periods are also completed by a noble gas.

The next period comprises the eight elements from lithium to neon; it is generally known as the *first short period*, or simply the *first row*. In this period the $2s$ and then the three $2p$ orbitals are consecutively occupied. In atoms having partially filled $2p$ orbitals the electrons prefer to maintain their spins mutually parallel, that is to have the same spin quantum numbers whenever possible. For this reason nitrogen has the electron configuration $1s^2 2s^2 2p_x^1 2p_y^1 2p_z^1$, which can conveniently be represented as

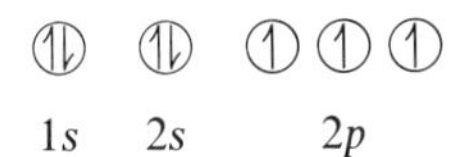

$1s$ $2s$ $2p$

where each circle represents an orbital and each arrow an electron, with the direction of the arrow indicating the orientation of the electron spin.

During the second short period, which comprises the eight elements from sodium to argon, the $3s$ and $3p$ orbitals are occupied. The $3d$ orbitals do not penetrate the *electron core*† to any great extent and are significantly higher in energy than the $3s$ and $3p$ orbitals. In fact the $4s$ orbital is lower in energy than the $3d$ and is the next orbital to be occupied.

The long periods

As predicted the next period begins with the filling of the $4s$ orbital, but then an important difference from the sequence of the short periods arises since the $3d$ orbitals are occupied before the $4p$. The ten elements beginning with scandium make up the first *transition series* of elements as the five $3d$ orbitals are filled. The first long period is completed by the six elements from gallium to krypton as the $4p$ orbitals are occupied, so that altogether there are eighteen elements in the period.

We might expect these transition elements to have the electron configuration (Ar core) $3d^x4s^2$ where $x = 0$ to 10. This is generally the case although, exceptionally, chromium has the ground state configuration (Ar core) $3d^54s^1$ and copper (Ar core) $3d^{10}4s^1$. These exceptions indicate the added stability of filled or half-filled shells of d electrons, and also show how close in energy are the $3d$ and $4s$ orbitals. We shall see later that the $3d$ electrons are stabilized with respect to the $4s$ on crossing the transition series, so that by the end of the series the $3d$ orbitals have entered the electron core and are no longer involved in chemical bonding.

The structure of the second long period is wholly analogous to that of the first, as the consecutive occupation of the $5s$, $4d$, and $5p$ orbitals gives the eighteen elements from rubidium to xenon, but the third long period is considerably more complex. It begins with the occupation of the $6s$ orbital giving the elements caesium and barium. Lanthanum is the next element; it has the ground state electron configuration (Xe core) $5d^16s^2$, and we might expect the third transition series to follow. However the seven $4f$ orbitals are occupied next, giving rise to a series of fourteen elements known as the *lanthanides*. It is not until the $4f$ orbitals are completely filled that further electrons enter the $5d$ orbitals, thus completing the third transition series; the period is completed with the occupation of the $6p$ orbitals giving the six elements from thallium to radon. This is the longest period and contains thirty-two elements in all.

The final period begins with the elements francium, radium, and actinium which have the electron configurations (Rn core) $7s^1$, $7s^2$, and $6d^17s^2$ respectively. The $5f$ orbitals are then filled giving a further series of fourteen elements from thorium to lawrencium known as the *actinide* elements. Further elements would be expected to form a fourth transition series but they have yet to be discovered. All the elements of this period are radioactive;

† The *electron core* comprises all electrons in an atom which are so tightly bound to the nucleus that they take no part in chemical combination. At argon all of the electrons are core electrons and the configuration $1s^22s^22p^63s^23p^6$ will in future be written in shorthand form as (Ar core).

many are very short-lived and have been synthesized in only minute quantities. For these reasons little is known about the chemistry of several of the elements.

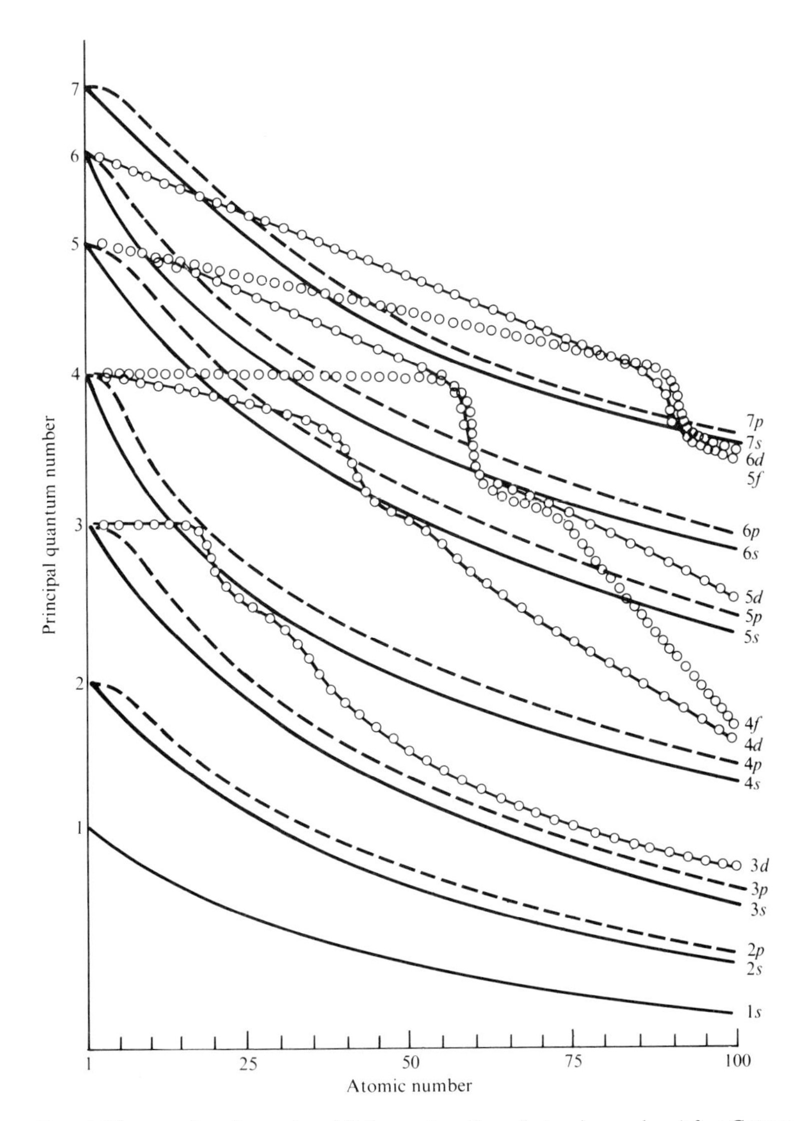

FIG. 6. The energies of atomic orbitals as a function of atomic number (after Cotton and Wilkinson, 1962).

Stability of atomic orbitals in the long periods

We saw that during the first transition series the 3*d* orbitals slowly became more stable than the 4*s*. A related effect is observed during subsequent long periods. Thus on crossing the lanthanide series the 4*f* orbitals are stabilized with respect to the 5*d* and 6*s* orbitals, so that by the end of the series they have entered the electron core. The overall variation of the stability of atomic orbitals with atomic number (the total number of electrons in an atom) is shown in Fig. 6. It shows clearly how the *d* and *f* orbitals change rapidly during the transition series from being high-energy orbitals to stable core orbitals.

Classification into groups

The classification of the elements into groups is shown in Fig. 5. In the short periods the elements are conveniently classified into groups I–VII with the noble gases as group 0†. As we shall see later (p. 24) the group number is, with some exceptions, the same as the maximum oxidation state‡ shown by the elements of the group. Elements within a group generally show strong similarities in their chemical behaviour.

In the long periods a further classification of the elements as A or B elements is made. Thus in the first long period both titanium and germanium have a maximum oxidation state of four and fall into groups IVA and IVB respectively. In general the pre-transition elements and early transition elements are labelled A elements, while the late transition elements and post-transition elements are B elements.

As can be seen from Fig. 5, the elements iron, cobalt, and nickel do not fall naturally into either classification. These elements and their analogues in subsequent periods are placed in group VIII after the first classification by Mendeleev, though few of the elements ever achieve an oxidation state of eight and differences in chemistry between the elements iron, cobalt, and nickel are greater than the similarities.

The elements may also be classified more directly according to the orbitals being filled. Thus the elements of groups IA and IIA are collectively called the *s*-block elements, those of groups IIIB–VIIB are the *p*-block, the transition metals are the *d*-block, and the lanthanides and actinides are the *f*-block elements.

PROBLEM

3.1. Predict the ground-state electron configurations of elements having the atomic numbers (the number of protons in the nucleus) 4, 16, and 22. Identify the elements and classify them according to (a) the group and (b) the block of the periodic table into which they fall.

† Strictly this group should be called group VIII but group VIII was used by Mendeleev to classify some transition elements (see p. 1).

‡ The term 'oxidation state' is discussed on p. 25.

4. Periodicity and the Atomic and Bonding Properties of the Main-Group Elements

HAVING established the periodic table of the elements, the remainder of this book is devoted to developing an understanding of the physical and chemical properties of the elements in terms of their positions in the periodic table. In this chapter we shall discuss trends in certain physical properties of the elements which can be used as a basis for discussing chemical behaviour.

Ionization energies of the elements

The *ionization energy* of an atom is the minimum energy required to remove an electron from an atom in its ground-state electron configuration. So for the magnesium atom the first ionization energy† is the energy required for the process

$$Mg \rightarrow Mg^{+} + e^{-}, \qquad I = 7{\cdot}4 \times 10^{5}\ J\ mol^{-1}.$$

Similarly the second ionization energy is the energy required to remove a second electron,

$$Mg^{+} \rightarrow Mg^{2+} + e^{-}, \qquad I = 14{\cdot}5 \times 10^{5}\ J\ mol^{-1}.$$

As might be expected, it is more difficult to remove an electron from a positively charged ion (cation) than from a neutral atom.

The first ionization energies of the elements of the short periods are plotted in Fig. 7. It is immediately apparent that the first ionization energy for lithium is much lower than that for helium. This is because the single $2s$ electron in lithium 'feels' a low effective nuclear charge since it is strongly shielded by the helium core $1s^2$ electrons; it is therefore easily removed. There is another drop in ionization energy between beryllium and boron. The low ionization energy for boron can be explained in terms of its electron configuration (He core) $2s^2 2p^1$; the $2p$ electron is shielded from the nucleus by the electrons in the filled $2s$ orbital and is readily lost. In subsequent elements the electrons in the $2p$ orbitals shield each other rather ineffectively from the nuclear charge so that the ionization energy increases as the $2p$ orbitals are filled‡, until a maximum value is reached at the noble gas neon. There is another break at the first element of the next period, sodium, in which the single $3s$ electron is effectively shielded by the neon core electrons.

In general then, the ionization energies increase irregularly on crossing a period.

† The ionization energy, I, is usually defined for one mol of atoms and is measured in Joules.

‡ Fig. 7 shows that the increase is not regular; there is a drop after nitrogen, which has the half-filled shell configuration $1s^2 2s^2 2p^3$.

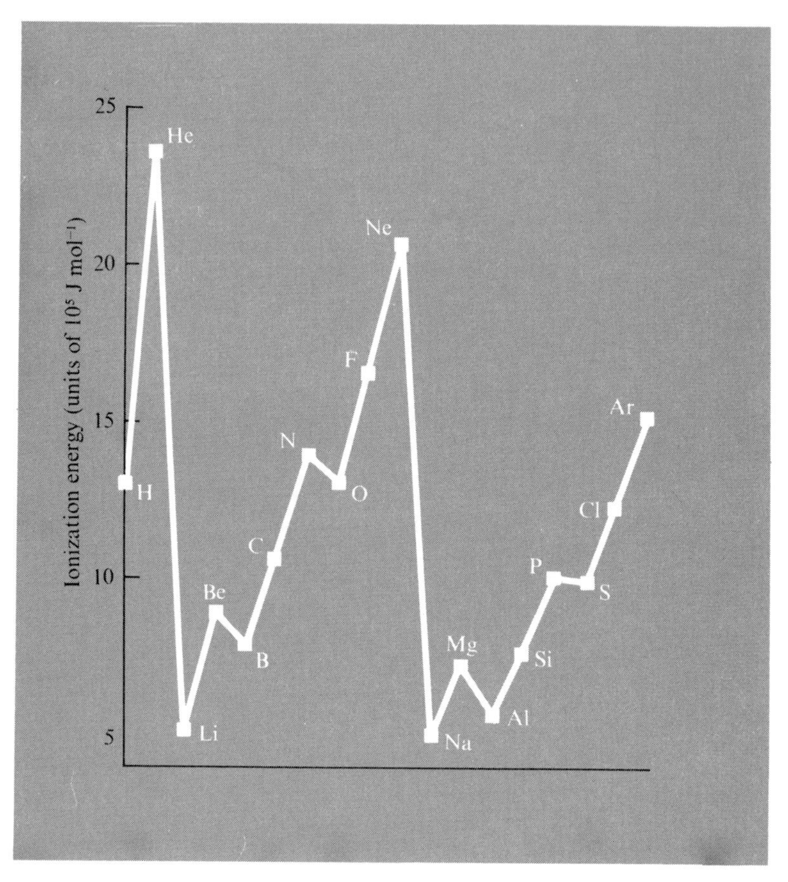

FIG. 7. Ionization energies for the elements of the short periods.

Comparison of the ionization energies within a group is also instructive, as shown in Fig. 8, which contains a plot of the ionization energies for the elements of groups IA and IB. The steady decrease in ionization energies in group IA (lithium to caesium) is due to the increasing *size* of the atoms on descending the group. The single 6*s* electron in the large caesium atom is much more readily removed than the 2*s* electron in lithium since the electrostatic attraction of the electron to the nucleus decreases with distance.

For the group IB elements the ionization energies are considerably higher, and there is no general decrease on descending the group. The ionization process for copper is

$$\mathrm{Cu}[(\mathrm{Ar\ core})\,3d^{10}4s^{1}] \rightarrow \mathrm{Cu}^{+}[(\mathrm{Ar\ core})\,3d^{10}] + \mathrm{e}^{-}.$$

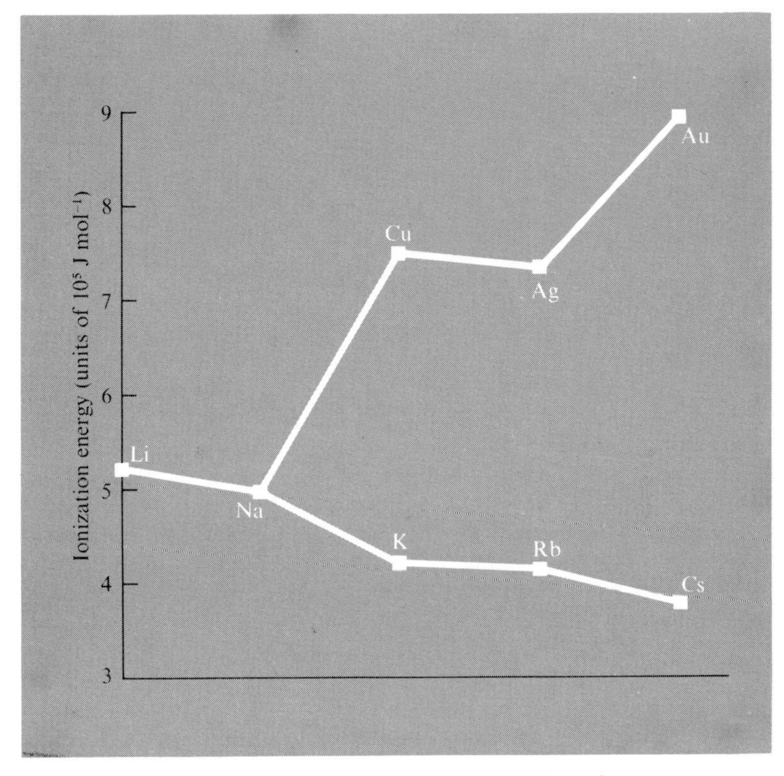

FIG. 8. Ionization energies for the group IA and IB elements.

The 4s electron is shielded by the argon core electrons and the filled subshell of d electrons. Now the strongly directional and diffuse d electrons shield the 4s electron rather ineffectively from the nuclear charge; it therefore 'feels' a high effective nuclear charge and the ionization energy is much higher than that of potassium, which has the configuration (Ar core) $4s^1$.

The ionization energy for the larger silver atom is, as expected, somewhat less than that for copper, but there is a large increase for gold. The ionization reaction for gold is

$$(\text{Xe core})\,4f^{14}5d^{10}6s^1 \rightarrow (\text{Xe core})\,4f^{14}5d^{10} + e^-.$$

The 6s electron is poorly shielded by both 4f and 5d electrons and a high ionization energy results. Relatively high ionization energies for the B elements and particularly for the post-lanthanide elements represent a general phenomenon which has important consequences for the chemistry of these elements.

Some chemical deductions can be made from ionization energy comparisons. Thus we expect that gaseous cations will be formed most readily by the group IA elements which have the lowest ionization energies, less readily by the group IB elements, and least readily by the group 0 noble gases. Second and subsequent ionization energies are also important. Thus Na^+ is more stable than Mg^+ with respect to their atoms; however, compare the ionization energies

$$Na^+ \text{ (neon core)} \rightarrow Na^{2+} \ (1s^2 2s^2 2p^5), \qquad I = 45{\cdot}6 \times 10^5 \text{ J mol}^{-1};$$

$$Mg^+ \text{ (neon core } 3s^1) \rightarrow Mg^{2+} \text{ (neon core)}, \qquad I = 14{\cdot}5 \times 10^5 \text{ J mol}^{-1}.$$

The ion Mg^{2+} is important in chemical compounds but Na^{2+} is never formed.

Electron affinities of the elements

The electron affinity of an atom is the energy *released* when an electron is added to the atom. It is therefore a measure of the ability of an atom to form a gaseous anion. There are great practical difficulties in measuring this property and so it is usually estimated by indirect methods; accurate values are known only for a few elements. We shall not discuss electron affinities at length, but mention only that the greatest values are found for the halogen atoms which, by gaining an electron, achieve noble-gas electron configurations.

Electronegativities of the elements

The electronegativity of an element is a measure of the power of the element to attract electrons to itself *in chemical compounds*†. We can illustrate this property by considering the hydrogen chloride molecule; the hydrogen–chlorine bond is intermediate in character between a covalent and an ionic bond:

H–Cl	and	H^+Cl^-.
(electron pair equally shared)		(bonding electrons all on chlorine)

Chlorine is the more electronegative element and so accumulates a net negative charge. We can represent hydrogen chloride as $H^{\delta+}$—$Cl^{\delta-}$ and write the overall wave function

$$\psi = \psi \text{ covalent} + c\psi \text{ ionic}$$

where c is a measure of the *ionic character* in the H–Cl bond and is determined by the difference in electronegativity between hydrogen and chlorine.

It is difficult to assign accurate electronegativities to the elements, partly because the precise values may vary in different compounds, but the concept

† It is thus different from the electron affinity which is a property of isolated atoms.

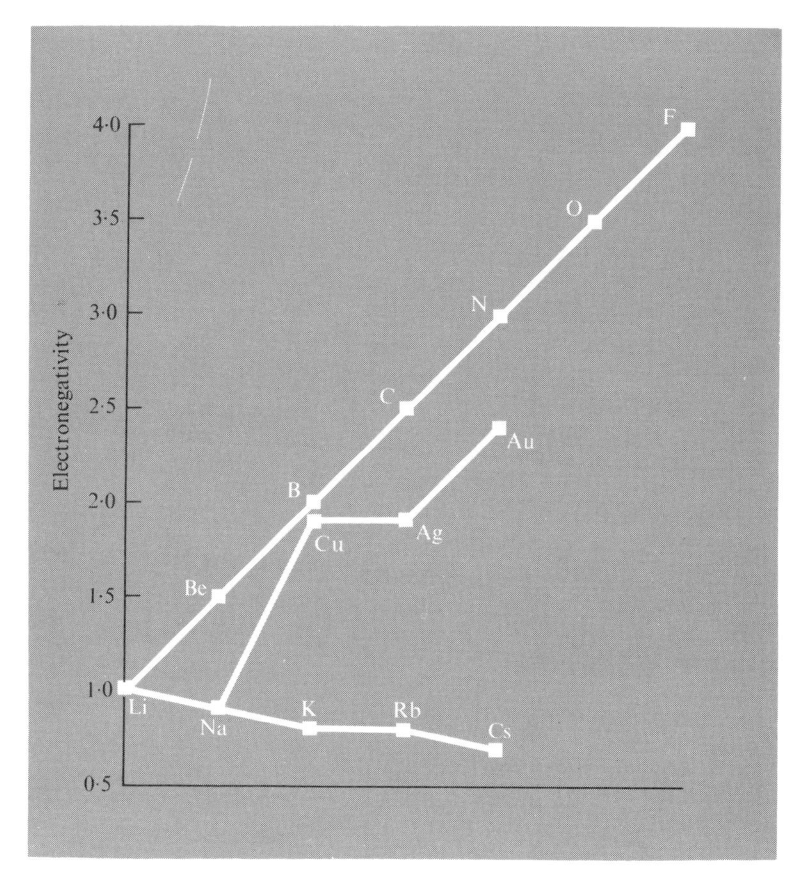

FIG. 9. Electronegativities of some elements.

is nevertheless extremely valuable. One quantitative estimate of electronegativity is the average of the ionization energy and the electron affinity for an element, that is the average of the ability of an atom of the element to donate or accept an electron.

Some electronegativity values for the elements of the first short period and for the group IA and IB elements are shown in Fig. 9. The trends are similar to those in the ionization energies shown in Figs. 7 and 8. Thus the electronegativities increase steadily on crossing a period and generally decrease on descending a group, though the latter trend may be reversed for the B elements.

It can be estimated that the bond between two elements will be essentially ionic in character if the electronegativity difference between the elements is

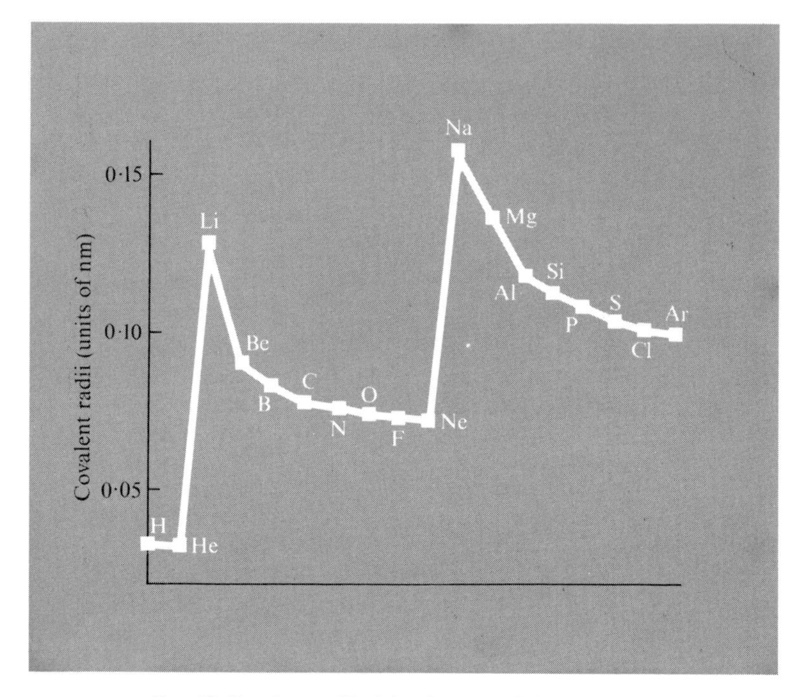

FIG. 10. Covalent radii of the elements of the short periods.

greater than two. We could correctly predict an ionic bond between lithium and fluorine (electronegativity difference 3·0) and a largely covalent bond between carbon and chlorine (electronegativity difference 0·5).

Covalent radii of the elements

The nature and value of the concept of covalent radius is best illustrated by an example. Thus the single bond covalent radius for carbon is defined as half the C—C bond distance of 0·154 nm† in ethane, CH_3—CH_3, and that for fluorine as half the F—F distance of 0·142 nm in the fluorine molecule. We can then use these covalent radii to compute the expected C—F distance in methyl fluoride, CH_3—F, to be (0·077 + 0·071) = 0·148 nm. The observed and calculated bond distances usually agree well, though, for the above example, the observed C—F distance is in fact 0·138 nm. This contraction from the expected value is a result of the extra bond strength arising from the partial ionic character of the C—F bond; by taking account of this factor, excellent agreement between calculated and observed covalent bond lengths may be

† 1nm = 10^{-9}m.

obtained. The covalent radii may be used directly to calculate bond lengths when there is little or no ionic character in the bonds.

Periodic trends in covalent radii can be correlated with similar trends in ionization energies. Thus a high effective nuclear charge leads to a high ionization energy and to a low covalent radius as the outer electrons are tightly held and so are pulled in closer to the nucleus. Covalent radii for the elements of the short periods are given in Fig. 10. Covalent radii decrease on crossing a period, increase on descending a group and are greater for the A elements than for the B elements.

Ionic radii of the elements

We have seen how covalent radii can be estimated by examining bond lengths in covalent compounds. By a similar, though slightly more complex, process, ionic radii can be obtained from internuclear distances in ionic compounds.

Figure 11 shows the covalent radii for the group IVB elements together with the ionic radii for the M^{2+} and M^{4+} ions; it also shows the covalent radii of the halogens (group VIIB elements) and the ionic radii of the halide ions. The radii *decrease rapidly when the ions carry a positive charge* and *increase with negative charge*; they increase on descending a group. We can readily explain these trends in terms of the effective nuclear charge felt by the outer electrons.

Polarizing power and polarizability

When a cation M^+ approaches an anion X^- it will tend to distort the electron cloud of the anion towards itself.

This *polarization* will lead to a build-up of electron charge between the nuclei and hence to some covalency in the bond. Now the importance of this covalency will depend on the *polarizing power* of the cation and the *polarizability* of the electron cloud of the anion. One aspect of polarizability is therefore an alternative to the use of electronegativity to predict the importance of covalency in a compound.

The polarizing power of a cation can be measured approximately by the ratio of its charge to its ionic radius, and is greatest for small, highly charged cations. Hence covalency will be more important in the series $B^{3+} > Be^{2+} > Li^+$ and $Li^+ > Na^+ > K^+$.

Similarly, large, highly charged anions are more readily polarized than smaller anions, so we expect covalency to be more important in the series

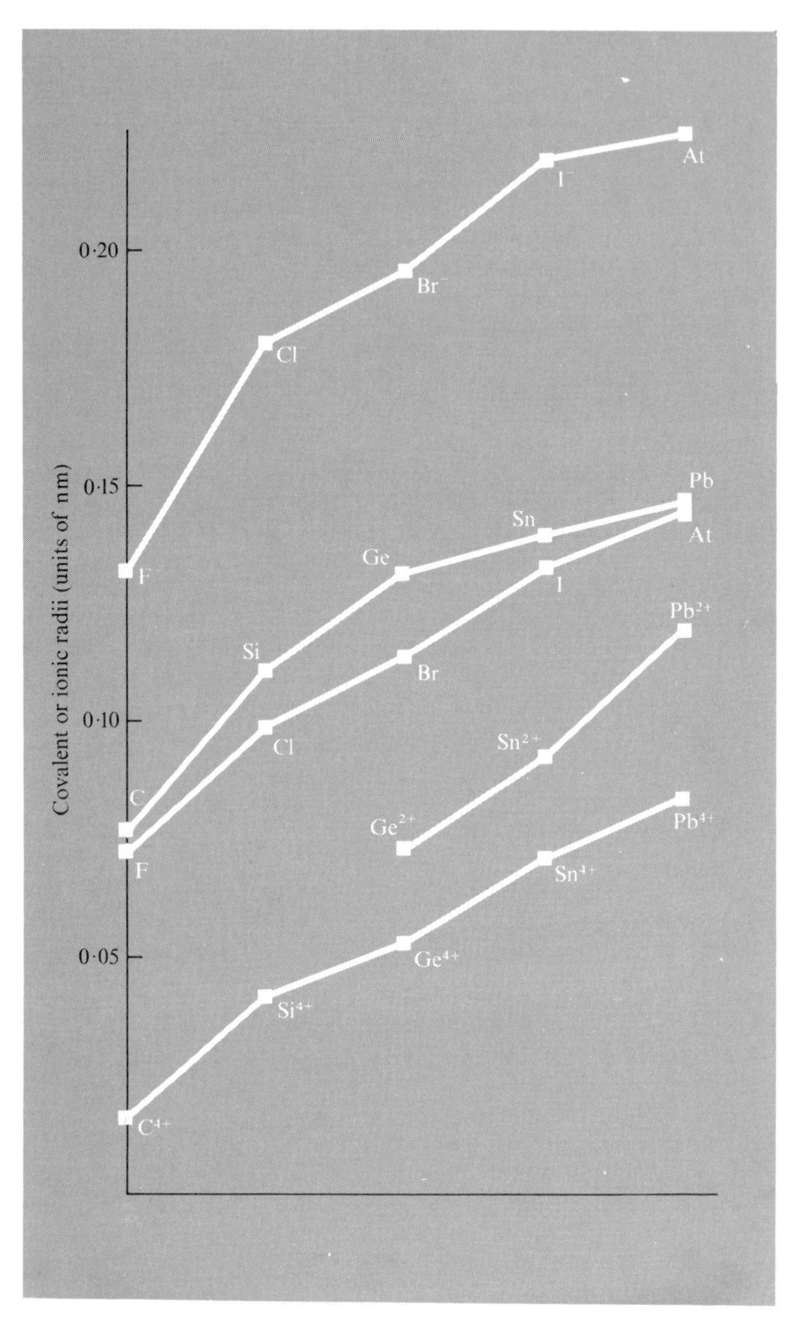

FIG. 11. Covalent radii and ionic radii for some element atoms or ions. (The values for C^{4+} and Si^{4+} are estimated, since no compounds containing these ions exist.)

$P^{3-} > S^{2-} > Cl^{-}$ and $I^{-} > Br^{-} > Cl^{-} > F^{-}$. To be more specific, we expect greater covalency in compounds like boron sulphide, B_2S_3, than in potassium fluoride, KF.

Ionic compounds

The lowest ionization energy of any element (that for caesium) is greater than the highest electron affinity (that for chlorine), so that even the atomic reaction

$$Cs + Cl \rightarrow Cs^{+} + Cl^{-}$$

would not take place if isolated ions were formed. In fact ionic *compounds* exist as regular arrays of cations and anions in solid lattices. The ionic lattice for caesium chloride is shown in Fig. 12. Each caesium ion is surrounded by eight chloride ions†; it is attracted to these ions electrostatically, but the electron shells produce a counter-repulsion. The sum over the whole lattice of the energy arising from these attractive and repulsive forces is called the *lattice energy*; it is this energy which is responsible for the stability of solid ionic compounds.

For a lattice containing cations of charge Z^{+} and anions of charge Z^{-} with a closest interionic separation of r, the lattice energy, U, is essentially

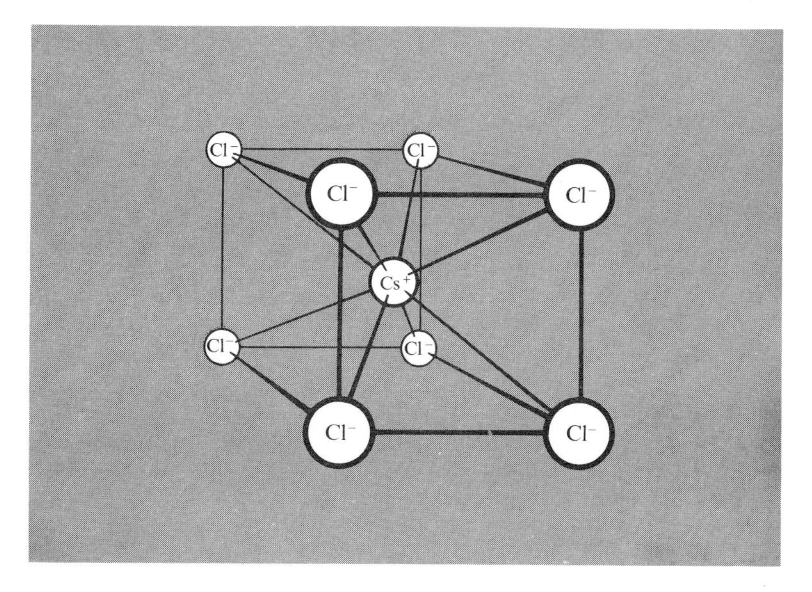

FIG. 12. Part of the lattice structure of caesium chloride.

† We say that the 'co-ordination number' of each caesium ion is eight.

proportional to Z^+Z^-/r. Hence lattice energies are greatest for small, highly charged cations and anions.

It is the balance between ionization energies and lattice energies that determines the stability of ionic compounds. For instance calcium forms the chloride $CaCl_2$ rather than CaCl or $CaCl_3$. The extra lattice energy for the Ca^{2+} ion more than balances the extra energy needed to form this ion from Ca^+, but to remove a further electron, which would come from the argon electron core, would require a prohibitive amount of energy. It is easy to see why the ionic compounds of the group I–III elements usually contain the elements in their group valency states.

Covalent compounds

The first modern theory of the covalent bond considered that bond formation took place by the sharing of electrons between two atoms giving an electron-pair bond, for instance

$$\mathrm{H}\cdot + \cdot\mathrm{Cl} \rightarrow \mathrm{H}\!:\!\mathrm{Cl}.$$

In more recent (valence bond) theory, the electron-pair bond is considered to be formed by overlap of atomic orbitals on the two atoms.

Let us consider formation of methane from carbon and hydrogen atoms. The ground-state electron configuration of carbon is

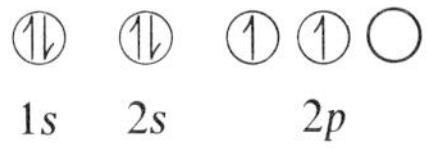

(↑↓) (↑↓) (↑)(↑)( )

1*s* 2*s* 2*p*

There are only two unpaired electrons so that only two covalent bonds can be formed. In order to form four bonds an electron must be 'promoted' from the 2*s* to the 2*p* orbital. The energy required is called the *promotion energy*. We now have the valence electron configuration at carbon:

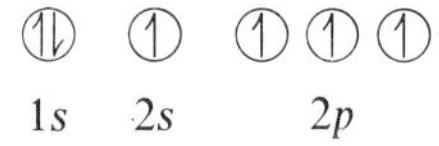

(↑↓) (↑) (↑)(↑)(↑)

1*s* 2*s* 2*p*

Each unpaired electron is then paired with the 1*s* electron of a hydrogen atom forming four electron-pair bonds. The stereochemistry of methane (the orientation of the atoms in the molecule) is governed by the need to minimize electrostatic repulsions between these bonding electron pairs. This is best achieved if the C–H bonds are directed towards the corners of a tetrahedron with the carbon atom at the centre and this is the observed stereochemistry of the methane molecule. We shall see that the stereochemistry of many simple covalent compounds can be deduced by similar considerations.

Co-ordination number and oxidation state

As well as the valency, the terms co-ordination number and oxidation state are used to describe the state of an atom in a chemical compound.

The co-ordination number is simply defined as the number of nearest neighbours to the given atom in a chemical compound.

The oxidation state of an atom in a compound is defined as the charge (with the sign) which it would carry in the most probable ionic formulation of the compound. This is decided on the basis of the electronegativities of the elements in the compound. Take for example the covalent compound $SnCl_4$. The most probable ionic formulation would be $Sn^{4+}Cl_4^-$, so that the oxidation state for tin is IV and that for chlorine is $-$I. The compound may be written as tin(IV)chloride.

The co-ordination number of an atom often differs from its oxidation state. Thus in the ion BF_4^- the oxidation state for boron in III but the co-ordination number is four, while in the ionic caesium chloride (p. 23) the oxidation state for caesium is I but its co-ordination number is eight.

PROBLEMS

4.1. Arrange the following groups of elements in order of
 i) increasing first ionization energy,
 ii) increasing electronegativity.

(a) B, N, F, Ne (b) Mg, Ca, Sr, Ba, Hg

4.2. Arrange the following ions in order of decreasing ionic radius. Which ion will be formed least readily from its atom?

Li^+, Be^{2+}, B^{3+}, Na^+

4.3. Give the oxidation state and co-ordination number of the central atom in each of the following molecules or ions.

$SiCl_4$, $SiCl_6^{2-}$, $AlCl_4^-$, AlF_6^{3-}

5. Periodicity and the Physical and Structural Properties of the Main-Group Elements

AN understanding of the physical properties of the elements (solid, liquid, or gaseous) requires some knowledge of their structures (if solid or liquid) and, more importantly, the magnitudes of the interatomic forces. The latter can be expressed in terms of the elemental binding energy, which is the extra energy (or stabilization) gained by an element in going to its usual state from the free atoms.

Binding energies of the elements

The binding energies for the elements of the short periods are shown in Fig. 13. On crossing a period they increase with the number of valence (outer)

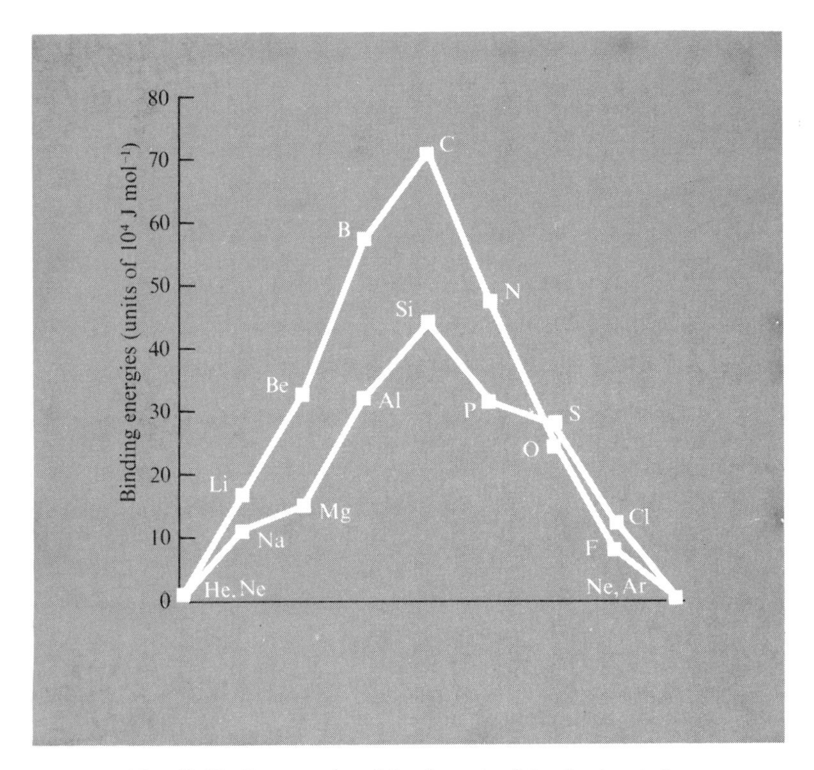

FIG. 13. Binding energies of the elements of the short periods.

electrons up to group IV and then fall to zero at the next noble gas. The noble gases are defined as having zero elemental binding energies simply because they exist as free atoms. At the other extreme carbon has the maximum number of bonding electrons. Thus in diamond each carbon atom is bound to four other carbon atoms in an infinite lattice; hence all four valence electrons on each carbon atom are engaged in bonding and a very high binding energy results. In comparison, boron has only three valence electrons. Nitrogen has five, but only three are engaged in forming the N≡N triple bond of the nitrogen molecule. As might be expected, boron and nitrogen have similar binding energies and each is lower than that for carbon. In the elements lithium and fluorine, each atom supplies only one bonding electron and the binding energies are low.

In general the elements with the greatest binding energies are unreactive while those with low binding energies are the most reactive. (The most notable exceptions to this rule are the noble gases which have zero binding energies but are most unreactive; this is readily explained in terms of the electron configurations of their atoms.) In order to gain further insight into the properties of other elements we must consider their structures in some detail.

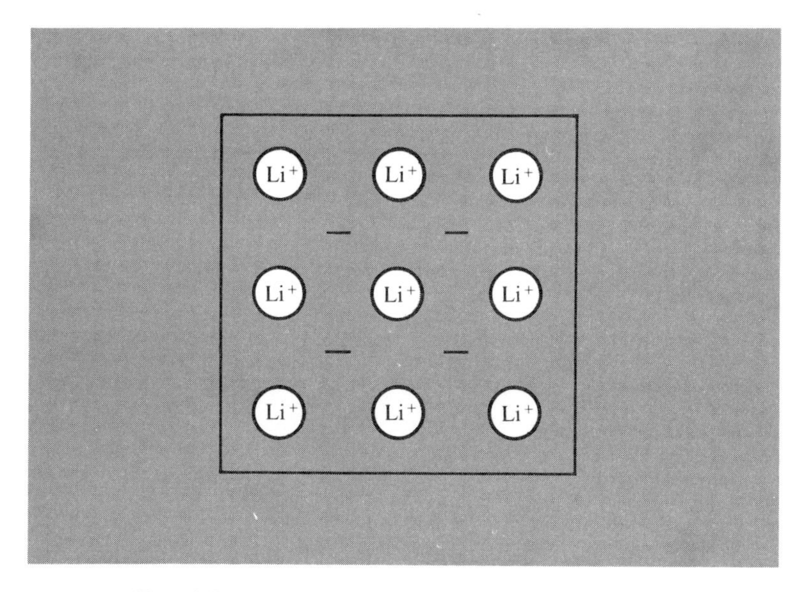

FIG. 14. Representation of the bonding in metallic lithium.

Elements of groups I and II

The elements of groups I and II are all metals; they have the typical metallic lustre and ability to conduct heat and electricity. These properties can be interpreted in terms of the structure and bonding in typical metals. As an example, we consider lithium metal to consist of a lattice of Li^+ ions with an equal number of electrons in a *valence band* which is delocalized over the entire lattice, as illustrated in Fig. 14.

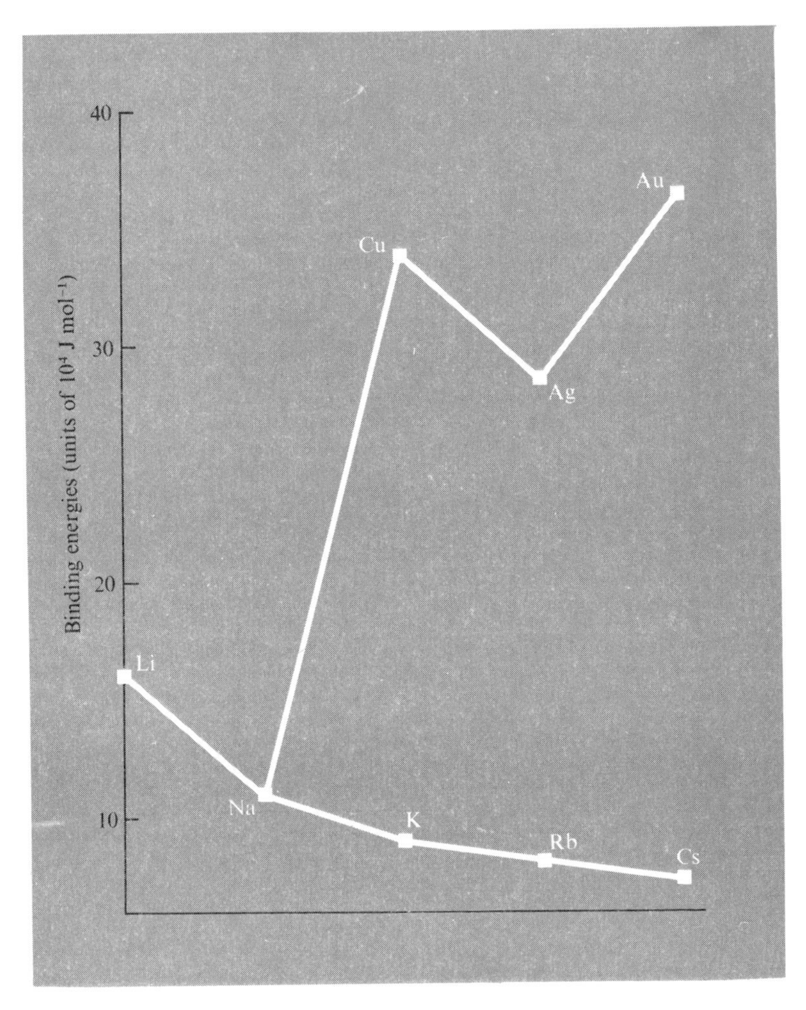

FIG. 15. Binding energies of the group IA and IB elements.

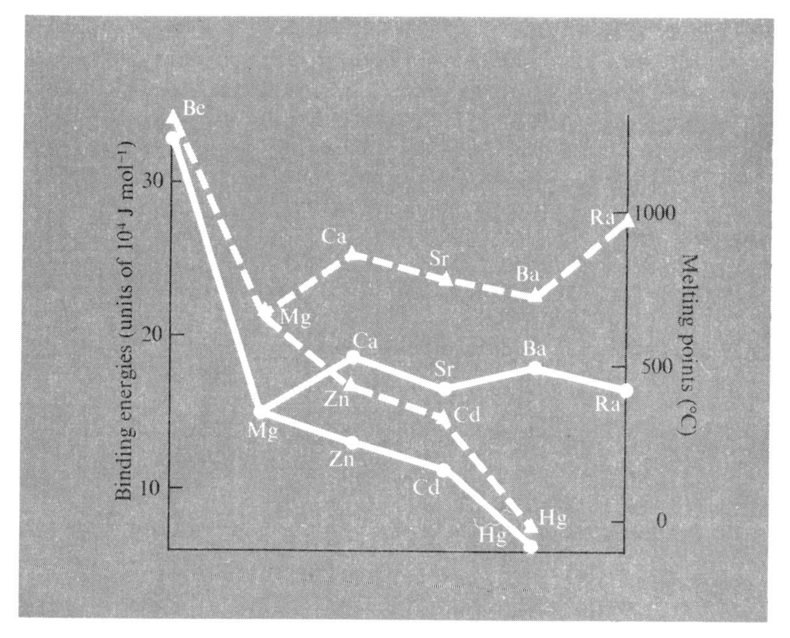

FIG. 16. Binding energies (O———) and melting points (△- - - -) of the group IIA and IIB elements.

The electrons can move freely throughout the lattice in the valence band and so bind each lithium ion to the other lithium ions in the lattice. This high electron mobility can also explain the high electrical and thermal conductivity shown by lithium and other metals.

For the group I elements the elemental binding energies are largely determined by the binding energies of the valence electrons in the individual atoms, that is by their ionization energies. This can be seen by comparing the plot of binding energies in Fig. 15 with that of ionization energies in Fig. 8.

We might expect the melting points and boiling points of the metals to correlate with their binding energies, since both melting and boiling involve, to some degree, atomization of the elements. That this is indeed the case can be seen from Fig. 16 which contains a plot of binding energies and melting points for the group II elements. Similarly both binding energies and melting points of the group IIA elements are considerably higher than those of the group IA elements.

Reactivity of metals

When the metallic elements react they normally lose electrons and form cations; for example

$$2Li + F_2 \rightarrow 2Li^+F^-.$$

A similar process takes place when lithium reacts with water or with oxygen. Thus the reactivity of the elements in these reactions is likely to depend on the ionization energy of the element atoms as well as on the elemental binding energy and the stability of the reaction products. This seems to be the case; thus the reactivity of the group IA elements increases on descending the group from Li → Cs as the ionization energy (and binding energy) decreases. Similarly the group IA elements are very much more reactive than the group IB elements, which have considerably higher ionization energies and binding energies. Thus caesium reacts explosively with water or dilute acids whereas gold does not react at all.

Elements of groups III and IV

The binding energies for boron and carbon are very high and the elements are therefore hard and chemically unreactive. The bonding in the solids may be interpreted in the same way as for the group I and II elements, but, since the ionization energies required to form the ions B^{3+} and C^{4+} are very high, the binding electrons are more localized and the elements are not metallic.

Boron is known as a *metalloid* because its properties are intermediate between those expected for a metal and for a non-metal. Thus it has a metallic lustre but is only a semiconductor of electricity. On descending group IIIB through the elements aluminium, gallium, indium, and thallium, the atoms increase in size and the binding energies and ionization energies decrease. The binding electrons then become more delocalized and have greater mobility so that the elements become increasingly metallic in character. Another consequence of the lower binding energies and ionization energies is that the heavier group IIIB elements are more reactive than boron towards water, oxygen, and the halogens, though they are considerably less reactive than the elements of groups IA and IIA.

A similar increase in metallic character is found on descending group IVB. Thus carbon has no metallic properties, silicon and germanium are metalloids having great importance as semiconductors of electricity, while tin and lead are more typical metals. The binding energies of all these elements are high, which partly explains the low chemical reactivity of the group IVB elements.

The elements carbon and tin can exist in more than one structural form, called allotropic forms. Thus carbon exists as diamond, the hardest known elemental substance, and as the much softer graphite which is a useful lubricant. Silicon, germanium, and one allotrope of tin have structures like that of diamond, while the other allotropes of tin and lead have more typical metallic structures. Because of these different structures, the trends in melting points and boiling points for the group IV elements are less regular than those we saw for the elements of groups I and II.

Elements of groups V and VI

There is a steady increase in metallic properties on descending these groups, as for groups III and IV. However an important difference from earlier groups arises, as the lighter elements of groups V and VI exist as small, covalent molecules rather than as three-dimensional lattices. For instance nitrogen exists as N_2 molecules, each containing a nitrogen–nitrogen triple bond. This bond is very strong so that nitrogen is an unreactive element, but although the binding energy is high the attractive forces between individual N_2 molecules (the intermolecular forces) are extremely weak. This has a profound effect on the physical properties of nitrogen which is a gas, and boils at $-197°C$ at atmospheric pressure.

The heavier elements do not form stable multiple bonds so that they exist as larger molecular units. Thus yellow phosphorus contains P_4 molecules

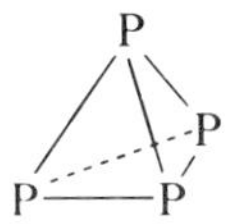

while the red and black allotropes contain much larger molecular units. Arsenic and antimony behave rather similarly, but bismuth exists only in a typical metallic lattice structure.

Similar trends are observed in group VI. Thus oxygen is a gas containing O_2 molecules with a double oxygen–oxygen bond, while sulphur typically contains cyclic S_8 molecules which are crown-shaped

Selenium and tellurium have chain structures and are semiconductors. The final element in group VI, polonium, is a more typical metal and is, for instance, a conductor of electricity.

Elements of group VII and group O

The group VIIB elements or halogens all consist of diatomic molecules with a single element–element bond, while the group 0 elements exist as the free atoms. The attractive forces between the halogen molecules or the noble gas atoms are largely determined by their *size*, and increase with increasing size of the atoms or molecules. In turn, the melting points and boiling points of the elements are determined by the magnitude of these interatomic or intermolecular forces (p. 29), and both of these physical properties increase in value on descending a group, as shown in Fig. 17.

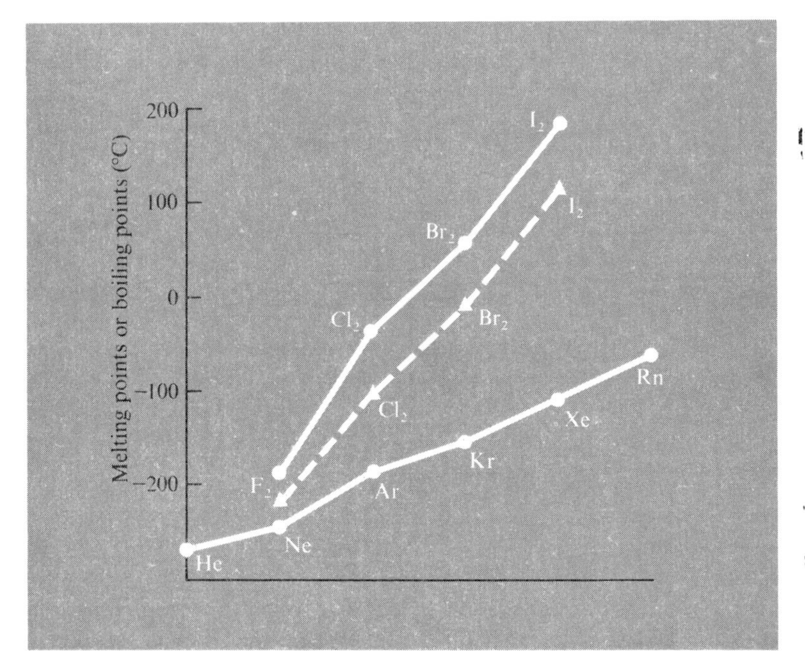

FIG. 17. Boiling points (○———) and melting points (△- - - -) of the group VIIB and group 0 elements.

None of the elements of these groups has any metallic properties. However while the group 0 elements are most unreactive, the group VIIB elements are very reactive. Fluorine is the most reactive of all elements and will react with almost every other element in the periodic table. As the group I elements tend to lose an electron and attain a noble-gas electron configuration, the fluorine atom tends to gain an electron, forming the fluoride ion, in order to

TABLE 1

The classification of the main-group elements.

Metals		*Metalloids*			*Non-metals*		
						H	He
Li	Be	B	C	N	O	F	Ne
Na	Mg	Al	Si	P	S	Cl	Ar
K	Ca	Ga	Ge	As	Se	Br	Kr
Rb	Sr	In	Sn	Sb	Te	I	Xe
Cs	Ba	Tl	Pb	Bi	Po	At	Rn

reach the noble-gas electron configuration of neon. We shall discuss the fluorides and other halides of the elements in Chapter 6.

Survey of the periodic table

The classification of the main-group elements as metals, metalloids, or non-metals is shown in Table 1. Bearing in mind that the transition elements and the lanthanide and actinide elements are all metals, it is clear that a high proportion of the chemical elements are metallic in nature and that this is especially pronounced for the heavier elements. Thus in the first short period of the periodic table there are two metals (Li and Be) and six other elements, while in the third long period there are thirty metals (Cs to Po) and only two non-metals (At and Rn).

PROBLEMS

5.1. Explain why boron and nitrogen have very different physical properties though their elemental binding energies are similar.

5.2. Arrange the following groups of elements in order of increasing melting point.
(a) Na, Mg, Al, Si, K (b) P, S, Cl, Ar
How do you explain these trends?

5.3. Arrange the following elements in order of their ability to conduct electricity.
(a) Al, Si, P (b) Se, Te, Po.

6. Periodicity and Chemical Compounds: The Main-Group Elements

THE chemical compounds of the transition metals and the lanthanides and actinides have certain characteristic properties because their ions have partially filled shells of *d* or *f* electrons. For this reason they are treated separately (in Chapters 7 and 8) from the main-group, or typical elements. Certain transition elements however do not show these characteristic properties. For instance zinc commonly forms a dipositive ion Zn^{2+} whose electron configuration is (Ar core) $3d^{10}$. The ion has a filled shell of *d* electrons and does not have characteristic transition-metal properties. For this reason the chemistry of zinc, and of some other elements which are formally transition elements, is included in this chapter.

In discussing the properties of chemical compounds we shall aim to treat classes of compounds which are formed by all elements of the periodic table. In this way periodic trends in the structure, bonding, and chemical properties of the compounds of the elements can be identified most readily.

We shall begin by discussing in some detail the properties of the halides of the elements, since these compounds illustrate very well the principles of chemical combination. Other classes of compounds can then be treated more briefly and with greater emphasis on trends in chemical behaviour.

Halides of the elements

Since the halogens are strongly electronegative elements (p. 19) their atoms normally carry some negative charge in their compounds with other elements; the magnitude of this charge depends on the electronegativity difference between the halogen and the particular element concerned. In *ionic halides* such as sodium chloride *all* of the bonding electrons are transferred to the halogen giving the halide ion.

$$2Na + Cl_2 \rightarrow 2Na^+Cl^-.$$

As we saw in Chapter 4, ionic compounds form lattice structures with strong interionic forces; they are therefore solids with high melting points and boiling points.

In *covalent halides* such as carbon tetrachloride the bonding electrons are localized in the carbon–chlorine bonds of individual CCl_4 molecules. The forces between these individual molecules—the intermolecular forces—are weak so that covalent halides are often gases, liquids, or low melting-point solids.

Thus the physical properties of element halides are largely determined by the nature of the bonding, and the classification into ionic or covalent halides

TABLE 2

The fluorides of the elements in their group oxidation states

Gp IA	Gp IIA	Gp IIB	Gp IIIB	Gp IVB	Gp VB	Gp VIB	Gp VIIB
LiF	BeF_2		BF_3	CF_4			
NaF	MgF_2		AlF_3	SiF_4	PF_5	SF_6	
KF	CaF_2	ZnF_2	GaF_3	GeF_4	AsF_5	SeF_6	
RbF	SrF_2	CdF_2	InF_3	SnF_4	SbF_5	TeF_6	IF_7
CsF	BaF_2	HgF_2	TlF_3	PbF_4	BiF_5		
	Ionic			*Polymeric*		*Covalent*	

is useful. A classification of the fluorides of the elements *in their group oxidation states* is shown in Table 2. Many of the elements form fluorides which do not readily fall into either ionic or covalent classification. These compounds often have polymeric structures in which molecules are associated through fluorine 'bridges'; they often have high melting points and boiling points but cannot accurately be described as ionic compounds, and are therefore classified separately as polymeric fluorides.

The usefulness of this classification is limited because the other halides are often very different from the fluorides, and because many of the elements form halides in oxidation states lower than the group number. However Table 2 does show some important trends. Thus it can be seen that on crossing the periodic table covalent compounds become more important, that the polymeric fluorides are generally formed by the heavier elements in a group, and that the first-row elements of groups V–VII do not form compounds in the group oxidation state. In order to explain these trends, and to compare the fluorides with the other halides of the elements, a more detailed survey of the properties of the element halides in groups is given in the following subsections.

Halides of the group IA elements

The group IA elements readily lose their single valence electron to form monopositive cations (p. 15), and all of the halides of these elements are typical ionic compounds. As expected for ionic compounds they are all high melting-point solids; the melting points and boiling points of the sodium halides are shown in Fig. 18. It can be seen that they decrease regularly as the size of the halide ion increases. We might have predicted this trend, for as the size of the halide ion increases the lattice energy decreases (p. 24). Since the lattice is partially or totally broken down when an ionic solid is melted or

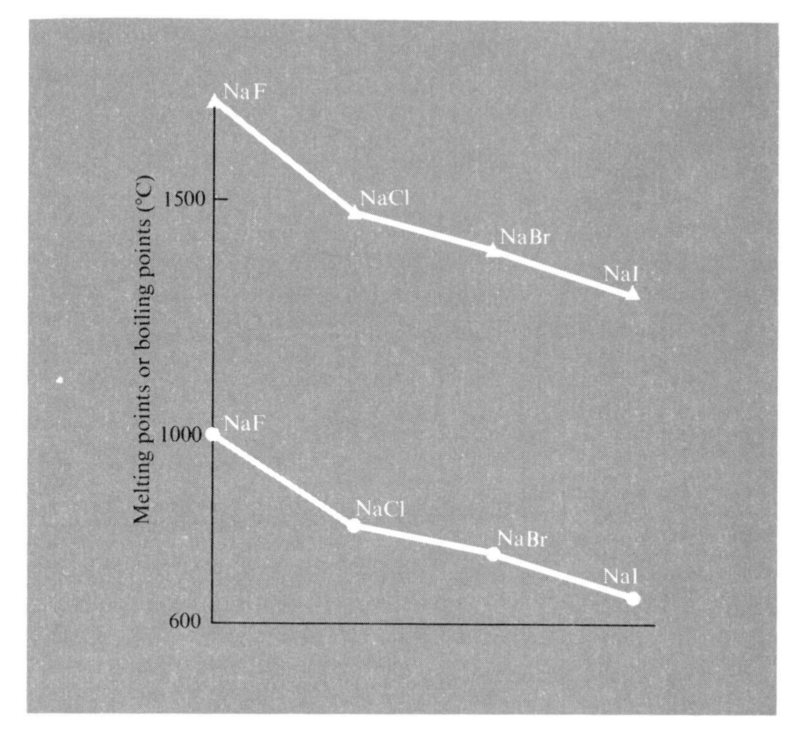

FIG. 18. Melting points (○———) and boiling points (△———) of the sodium halides.

boiled, the melting point and boiling point also decrease, as less energy is required to break down the lattice.

Halides of the group IIA and IIB elements

The group IIA metals have the electron configuration (noble gas core) ns^2. The extra energy needed to remove both valence electrons rather than just one is small compared with the extra lattice energy obtained by forming lattices with the dipositive cations such as Ba^{2+}, so that ionic halides MX_2 (for example $BaCl_2$ and $CaBr_2$) rather than MX are formed. Similarly the group IIB elements such as zinc, whose electron configuration is (Ar core) $3d^{10}4s^2$, lose both *ns* electrons and form essentially ionic halides such as $ZnBr_2$.

The melting points of the group II metal chlorides are shown in Table 3. In group IIA there is a steady *increase* in melting point as the cation size increases, whereas a steady decrease is predicted on the basis of ionic lattice

TABLE 3

Melting points of the group II metal chlorides (°C)

$MgCl_2$ 714	$CaCl_2$ 782	$SrCl_2$ 875	$BaCl_2$ 962
	$ZnCl_2$ 275	$CdCl_2$ 568	$HgCl_2$ 277

energies. The explanation of this effect is that the chlorides of the small cations such as Mg^{2+}, which have high polarizing power, are partially covalent and so have anomalously low melting points. The group IIB metal halides are also partially covalent and have low melting points. Thus mercury has the electron configuration (Xe core) $4f^{14}5d^{10}6s^2$, with the $6s$ electrons ineffectively shielded from the nuclear charge by the $4f$ and $5d$ electrons. Thus the ionization energy needed to form the Hg^{2+} ion is high and the mercury–halogen bonds in the mercuric halides have appreciable covalent character.

The beryllium halides are typical covalent compounds having none of the properties expected of ionic salts. The first element of group II therefore differs considerably from the other elements; in this case the difference can be attributed to the small size of the beryllium atom. Either the ionization energy needed to form the Be^{2+} ion is too great, or the high polarizing power of the Be^{2+} ion leads to bonding which is largely covalent. We shall see that the first elements in subsequent groups often have some unique properties not possessed by the heavier members of the group.

In order for beryllium to form two covalent electron-pair bonds it is necessary to promote an electron from the filled $2s$ to a vacant $2p$ orbital

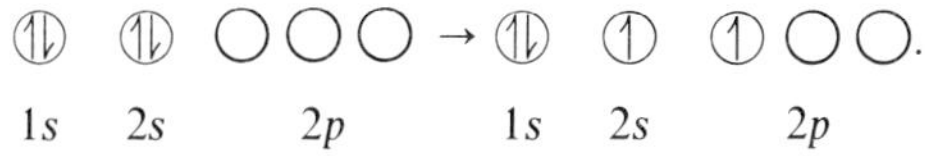

$$\underset{1s}{(\uparrow\downarrow)}\ \underset{2s}{(\uparrow\downarrow)}\ \underset{2p}{(\ )(\ )(\ )} \rightarrow \underset{1s}{(\uparrow\downarrow)}\ \underset{2s}{(\uparrow)}\ \underset{2p}{(\uparrow)(\ )(\ )}.$$

In order to minimize repulsions between bonding electron pairs we would expect the beryllium halides such as $BeCl_2$ to be linear molecules.

$$Cl—Be—Cl$$

However there is a strong tendency for the beryllium atom to accept two more electron pairs into the remaining vacant $2p$ orbitals, thus attaining the neon electron configuration. This is accomplished in beryllium chloride by the donation of a lone pair of electrons by each chlorine atom, giving the polymeric chain structure

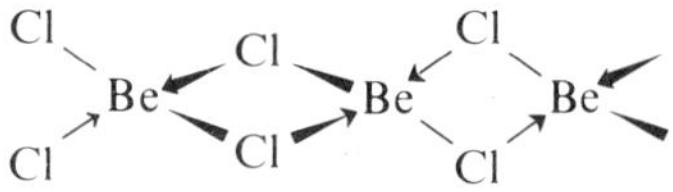

Cl Cl Cl
Be Be Be
Cl Cl Cl

where the notation Cl → Be represents an electron-pair bond in which both electrons are donated by the chlorine atom. The co-ordination number at each beryllium atom is four; this co-ordination number is very rarely exceeded for any element of the first short period.

Halides of the groups III and IV elements

The trend towards increasing covalency in the element–halogen bonds on crossing the periodic table continues in these groups, and typical halides such as BF_3 and $SiCl_4$ are covalent molecules.

In order to form the trihalides of the group III elements an electron must be promoted.

(core electrons) (↑↓) (↑)( )( ) → (↑) (↑)(↑)( ).

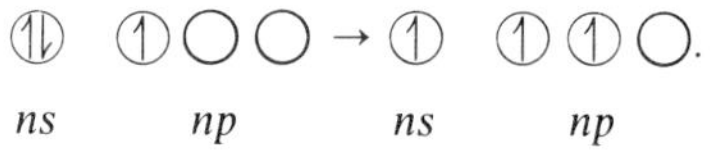

ns *np* *ns* *np*

The trihalides of aluminium, gallium, and indium dimerize, in the same way that beryllium chloride polymerizes, so that the metal atoms attain the noble-gas electron configuration. Thus aluminium chloride has the structure

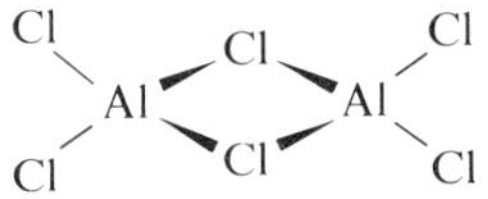

Cl Cl Cl
Al Al
Cl Cl Cl

The trifluorides of these elements are more strongly associated. Thus AlF_3 exists in an extended polymeric structure in which each aluminium atom is surrounded by six shared fluorine atoms. Hence aluminium attains a co-ordination number of six. The same co-ordination number is achieved when aluminium fluoride, AlF_3, reacts with three fluoride ions to give the ion AlF_6^{3-}:

$$AlF_3 + 3F^- \rightarrow AlF_6^{3-}.$$

We shall see that this co-ordination number of six is quite common, but is almost never found in the elements of the first short period, where the maximum co-ordination number, with very few exceptions, is four.

Exceptionally, the boron trihalides such as BF_3 do not dimerize or polymerize and have a trigonal planar structure:

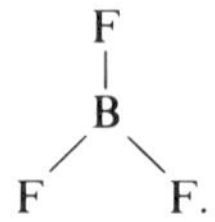

F
|
B
F F.

Boron trifluoride reacts with *one* fluoride ion to give the ion BF_4^- (the co-ordination number of boron is four—compare this with aluminium which forms AlF_6^{3-}).

In order for the group IV elements to form four covalent bonds it is again necessary to promote an *s* electron to a *p* orbital

$$\underset{ns}{(\uparrow\downarrow)} \quad \underset{np}{(\uparrow)(\uparrow)(\ )} \rightarrow \underset{ns}{(\uparrow)} \quad \underset{np}{(\uparrow)(\uparrow)(\uparrow)}.$$

The elements achieve the noble-gas electron configuration by forming the four covalent bonds, so that there is less tendency for the halides to dimerize or polymerize. For carbon the maximum co-ordination number is reached in the carbon tetrahalides and these are quite unreactive compounds. This is not true, however, for the tetrahalides of the other group IV elements, which have empty *d* orbitals capable of taking part in bonding. Thus silicon tetrachloride can accept electron pairs into its empty 3*d* orbitals and form ions such as $SiCl_5^-$ and $SiCl_6^{2-}$ in which the *co-ordination number* at silicon is five and six respectively.

This tendency to increase the co-ordination number above four is still greater for tin and lead, so that tin tetrafluoride (though not the other halides) polymerizes in order that the tin atoms may attain a co-ordination number of six.

Because of this polymeric structure, the melting point and boiling point for SnF_4 are considerably higher than for the other group IV tetrahalides, as shown in Fig. 19.

Lower oxidation-state halides. We have seen that in order for the group III and group IV elements to achieve their maximum oxidation state it is necessary to promote an electron from an *s* to a *p* orbital. The energy needed for this is usually more than offset by formation of two extra covalent bonds, but this may not always be the case. On descending a group the elements become larger and form weaker covalent bonds; gradually, the tendency to form the maximum oxidation state of three or four is decreased, and the tendency to form lower oxidation-state halides MX or MX_2 is increased. The latter tendency is greatest for the post-lanthanide elements thallium and lead for which the promotion energies are also high due to poor shielding of the 6*s* electrons. These elements form their most stable halides in an oxidation state two less than the group number, typical examples being TlBr and PbI_2. Formation of these lower oxidation-state compounds is referred to as the *inert pair* effect, since the ns^2 electron pair takes no part in bonding.

The higher oxidation-state compounds are most stable for the lightest halogens since these form the stronger covalent bonds. Thus TlF_3 is quite stable, but $TlCl_3$ decomposes readily to form TlCl and chlorine, while $TlBr_3$

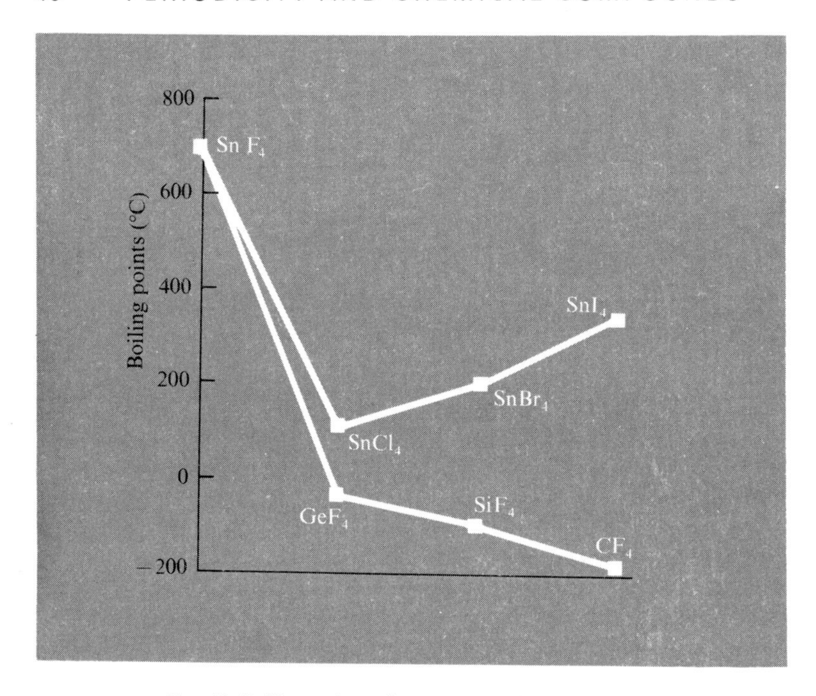

FIG. 19. Boiling points of some group IVB tetrahalides.

and TlI_3 are still less stable. Similarly PbF_4 is stable, $PbCl_4$ is unstable, and $PbBr_4$ and PbI_4 cannot be prepared.

The lighter elements indium and tin also form isolable halides such as InCl and $SnBr_2$, but these are considerably less stable than the corresponding derivatives of thallium and lead and are readily oxidized to indium (III) or tin (IV) derivatives.

The ions Tl^+ and Pb^{2+} have much lower polarizing power than Tl^{3+} and Pb^{4+} so that the lower oxidation-state halides are the more ionic. Thus thallium (I) chloride, TlCl, is a typical ionic compound and is similar in many ways to the halides of the group IA elements. The bonding in $PbCl_2$ and in $SnCl_2$ is intermediate between covalent and ionic and these compounds resemble mercuric chloride in many of their properties.

Halides of the group V and VI elements

The electronegativities of the elements in these groups approach those of the halogens so that all the halides are covalent compounds.

Nitrogen and oxygen have the respective valence electron configurations

N (↑↓) (↑)(↑)(↑)

O (↑↓) (↑↓)(↑)(↑).

2*s* 2*p*

Thus, without any electrons being promoted, nitrogen can form three and oxygen two covalent bonds. In each case there are four electron pairs in the valence electron shell so that NF_3 is pyramidal and OF_2 has a bent structure with *stereochemically active* lone pairs of electrons occupying the other tetrahedral positions.

Thus the oxidation state is not the same as the group number but is eight minus the group number in each case. The heavier elements of these groups also form similar compounds, MX_3 in group V or MX_2 in group VI, but, since they have vacant *d* orbitals which may take part in bonding, they may also form higher oxidation-state halides.

Thus, by promoting an electron from the 3*s* to a vacant 3*d* orbital phosphorus may reach a valence electron configuration

(↑↓) (↑)(↑)(↑) ( )( )( )( )( ) → (↑) (↑)(↑)(↑) (↑)( )( )( )( )

3*s* 3*p* 3*d* 3*s* 3*p* 3*d*

having five unpaired electrons, so that five covalent bonds may be formed. The pentahalides PF_5, PCl_5, and PBr_5 are known, but PI_5 cannot be prepared. The stability of the pentavalent compounds decreases on descending the group until for bismuth only the rather unstable BiF_5 can be prepared. The pentahalides may add a further halide ion to give a six co-ordinate anion.

$$PF_5 + F^- \rightarrow PF_6^-.$$

In group VI there are further possibilities. Thus for sulphur a single electron can be promoted from a 3*p* to a 3*d* orbital, allowing four covalent bonds to be formed.

(↑↓) (↑)(↑)(↑) (↑)( )( )( )( ).

3*s* 3*p* 3*d*

If a 3*s* electron is also promoted to a 3*d* orbital a covalency of six is possible.

(↑) (↑)(↑)(↑) (↑)(↑)( )( )( ).

3*s* 3*p* 3*d*

Examples of both types of compound are known and are more common than the simple two-covalent compounds. Thus sulphur forms SCl_4 and SF_6 as well as SCl_2, and similar compounds are formed by selenium and tellurium.

The formation of SF_6 is a further example of co-ordination number six for the elements of the second short period. On crossing the period, the iso-electronic series of ions AlF_6^{3-}, SiF_6^{2-}, PF_6^-, and SF_6 is known, while in the first short period co-ordination number four is more typical as in the ions BeF_4^{2-}, BF_4^-, and CF_4.

Halides of group VII elements

As well as the simple halogens like I_2, several *interhalogen* compounds such as ClF, BrCl, or IBr can be formed. The heavier elements, perhaps by using their vacant *d* orbitals, can also form higher oxidation-state compounds, especially in combination with fluorine. Thus chlorine forms ClF_3, bromine forms BrF_3 and BrF_5, and iodine forms IF_5 and IF_7. It will be noticed that the highest oxidation states are observed for the *heaviest* elements, in contrast to the trends observed in groups III–V. In this group the stability of higher oxidation-state compounds appears to depend on the promotion energies, which steadily decrease on descending the group, rather than on the covalent bond strength.

Halides of group 0 elements

Until recently it was thought that the noble gases would never form chemical compounds. However in 1962 the first such compound was isolated and several more have since been prepared. Thus xenon reacts directly with fluorine and XeF_2, XeF_4, and XeF_6 can be isolated. In order to form these compounds one or more electrons must be promoted from the 5*p* to the 5*d* orbitals of xenon. Like the ionization energy for xenon, this promotion energy is very high and stable compounds are only formed with the most powerful oxidizing agent fluorine. The promotion energies are even higher for the lighter noble gases so that argon forms no compounds with fluorine and krypton forms only KrF_2.

Reactions of halides with water

The group IA metal halides are mostly soluble in water. On dissolution the lattice energy of the crystalline halide is lost as the ions are separated, but extra energy called the solvation energy is gained by interaction of the ions with water. Since water, H_2O, is a polar molecule, it may interact with cations or anions as illustrated below for sodium and chloride ions.

$$Na^+ \cdots O^{\delta-}(H^{\delta+})_2 \qquad Cl^- \cdots H^{\delta+}-O^{\delta-}-H^{\delta+}$$

Sodium chloride is soluble in water partly because the energy gained by these electrostatic attractions is greater than the lattice energy which is lost.

The solvation energy for cations increases with their polarizing power, i.e. it increases with decrease in ionic size and with increase in charge. For the smaller group I cations (Li^+, Na^+) and all the group II metal cations the water molecules are tightly bound and may be retained when the halides are crystallized from aqueous solution. Thus beryllium bromide forms $BeBr_2{\cdot}4H_2O$ which contains the $Be(OH_2)_4^{2+}$ cation. Because of the stability of such hydrated salts, many anhydrous group II metal salts (for instance calcium chloride) are good drying agents.

Solutions of the halides of cations with still higher polarizing power are acidic because so much positive charge is developed on the hydrogen atoms of the attached water molecules that they may be lost as hydrogen ions. Thus aluminium chloride forms an acidic solution in water due to the reaction

$$AlCl_3 \xrightarrow{H_2O} Al(OH_2)_6^{3+} \rightleftharpoons Al(OH_2)_5OH^{2+} + H^+.$$

The hydrogen ions are in fact better represented as H_3O^+ rather than H^+ since the free proton has a very high polarizing power (being very small) and interacts strongly with water. It is for this reason that the covalent hydrogen chloride ionizes in water.

$$H_2O + HCl \rightarrow H_3O^+ + Cl^-.$$

Silicon tetrachloride gives a still more acidic solution than aluminium chloride, as it is completely hydrolysed by water to give silicic acid and hydrogen chloride†:

$$SiCl_4 \xrightarrow{H_2O} SiO_2{\cdot}xH_2O + HCl.$$

Phosphorus trichloride behaves like silicon tetrachloride, and elemental chlorine reacts with water to give a solution containing a mixture of hydrochloric and hypochlorous acids:

$$Cl_2 \xrightarrow{H_2O} HCl + HOCl.$$

Compounds of hydrogen with the elements

The known simple compounds of hydrogen with the main-group elements are shown in Table 4. They may be classified as shown in this table according to the nature of the bonding, which may be ionic, covalent, or intermediate between these two extremes.

Ionic hydrides

The hydrides of the group IA and the heavier group IIA elements are ionic and contain the H^- or hydride ion. The hydride ion is rather reactive,

† As noted on p. 39, CCl_4 is unreactive and is not affected by water.

TABLE 4

Classification of the element hydrides

Gp IA	Gp IIA	Gp IIB	Gp IIIB	Gp IVB	Gp VB	Gp VIB	Gp VIIB
LiH	BeH_2		BH_3	CH_4	NH_3	H_2O	HF
NaH	MgH_2		AlH_3	SiH_4	PH_3	H_2S	HCl
KH	CaH_2	ZnH_2	GaH_3	GeH_4	AsH_3	H_2Se	HBr
RbH	SrH_2	CdH_2	InH_3	SnH_4	SbH_3	H_2Te	HI
CsH	BaH_2			PbH_4	BiH_3		
	Ionic		*Intermediate*			*Covalent*	

so that ionic hydrides are reactive compounds. They react with water to give hydrogen according to the equation:

$$H^- + H_2O \rightarrow H_2 + OH^-.$$

Covalent hydrides

The covalent compounds of hydrogen are usually gases or volatile liquids. The types of compound formed are often similar to the halogen derivatives but the following important differences may be noted.

(1) Hydrogen never causes an expansion of the valence electron shell beyond the stable octet. Thus with sulphur, hydrogen forms H_2S like SCl_2, but not H_4S or H_6S, though SCl_4 and SF_6 are easily prepared.

(2) Since the electronegativity of hydrogen is considerably lower than that of fluorine, hydrogen may develop a net positive or negative charge in its compounds while fluorine *always* develops a negative charge. Thus while in SiH_4 the bond polarity is $Si^{\delta+}{-}H^{\delta-}$, and in CH_4 there is almost no polarity in the C—H bonds, in H_2S or HCl the polarity is $H^{\delta+}{-}S^{\delta-}$ or $H^{\delta+}{-}Cl^{\delta-}$. This difference in polarity is reflected in the different chemical properties of these compounds. Thus SiH_4 is slowly hydrolysed by water to give hydrogen (like the ionic hydrides), while H_2S and HCl are acidic compounds since they react with water to give hydrogen ions:

$$HCl + H_2O \rightarrow H_3O^+ + Cl^-.$$

The hydrides of group IIIB, such as AlH_3, may add a further hydride ion to give an anion:

$$AlH_3 + Li^+H^- \rightarrow Li^+AlH_4^-.$$

Advantage may be taken of the high reactivity of the ionic hydrides in the preparation of covalent hydrides such as tin (IV) hydride by the reaction

$$2CaH_2 + SnCl_4 \rightarrow 2CaCl_2 + SnH_4.$$

Lithium aluminium hydride, $LiAlH_4$, reacts similarly with covalent halides and is often used to prepare covalent hydrides.

The hydrogen bond

The melting and boiling points of the hydrogen compounds of the group IVB and group VIB elements are given in Fig. 20. It can be seen that, within a group, the boiling points increase on descending the group. The only exception is in group VIB where water has anomalously high melting and boiling points. The intermolecular forces between individual water molecules are greater than predicted, owing to the presence of *hydrogen bonding*. This effect is also present in ammonia, NH_3, and hydrogen fluoride, HF, which again have higher boiling points than might be expected. Thus, the hydrogen bond is only important when the hydrogen atom is bound to a small, electronegative

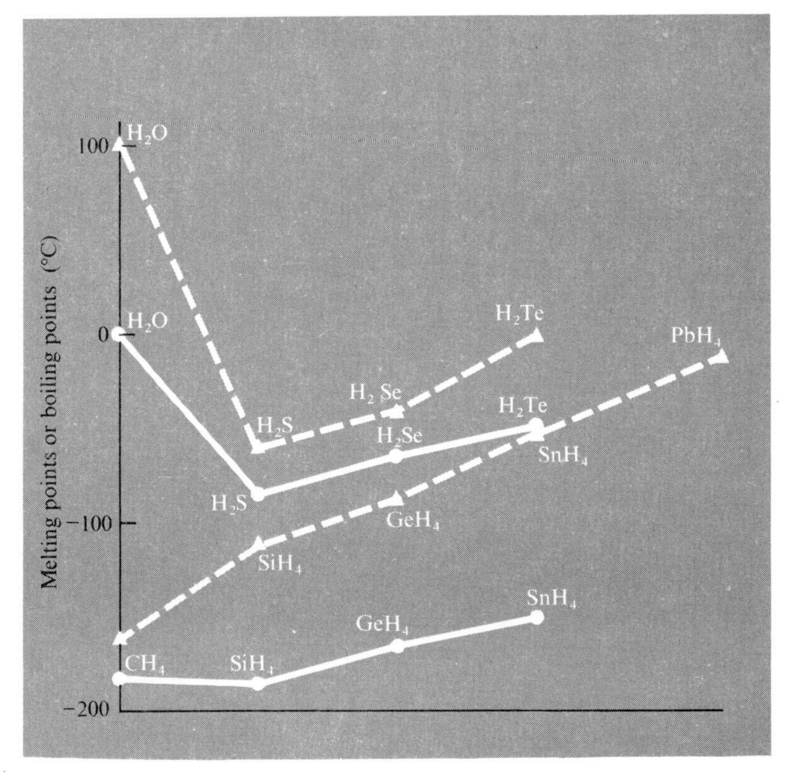

FIG. 20. Melting points (○———) and boiling points (△- - - -) of the compounds of hydrogen with the group IVB and VIB elements.

element. It is thought to be essentially an electrostatic interaction and is usually represented by a dotted line:

```
H
 \
  O···H
 /     \
H       O—H.
```

Methyl derivatives of the elements

These compounds form part of an important class of compounds in which an organic group such as the methyl, CH_3–, or ethyl, $CH_3{\cdot}CH_2$–, group is bound to another element. If this element is a metal the compounds are known as organometallic compounds. They have many resemblances to the metal hydrides.

Thus the methyl derivatives of the elements of groups IA and IIA such as methyl-lithium CH_3Li and dimethylmagnesium $(CH_3)_2Mg$ are high melting-point polymeric solids with partially ionic metal–carbon bonds, while those of the group VB–VIIB elements are usually gases or volatile liquids with covalent element–carbon bonds.

In the derivatives of the elements of groups I–III there is a polarity in the metal–carbon bond in the sense $M^{\delta+}—CH_3^{\delta-}$, and they react with water to give methane and the metal hydroxide (compare this with the reaction of ionic hydrides with water, p. 44):

$$CH_3Li + H_2O \rightarrow CH_4 + LiOH;$$

$$(CH_3)_3B + 3H_2O \rightarrow 3CH_4 + B(OH)_3.$$

In the compounds of the group IVB elements, such as tetramethyl lead, $(CH_3)_4Pb$, there is little polarity in the metal–carbon bonds, while the carbon atoms develop some positive charge in the methyl compounds of elements in subsequent groups such as trimethylphosphine, $(CH_3)_3P$, dimethyl sulphide, $(CH_3)_2S$, or methyl iodide, CH_3I. None of these compounds is affected by water under normal conditions.

The boiling points of the methyl derivatives of the elements of groups IVB and VIB are shown in Fig. 21. They increase steadily on descending each group (compare this with the behaviour of the element hydrides in Fig. 20, p. 45).

Catenation

Catenated compounds contain chains, rings, or clusters of identical atoms, like those we have already seen in the structures of the *elements*. Relatively few elements, however, form stable catenated *compounds*.

Catenation is most common in group IVB and is especially important for carbon, as the great variety of compounds containing chains of carbon atoms

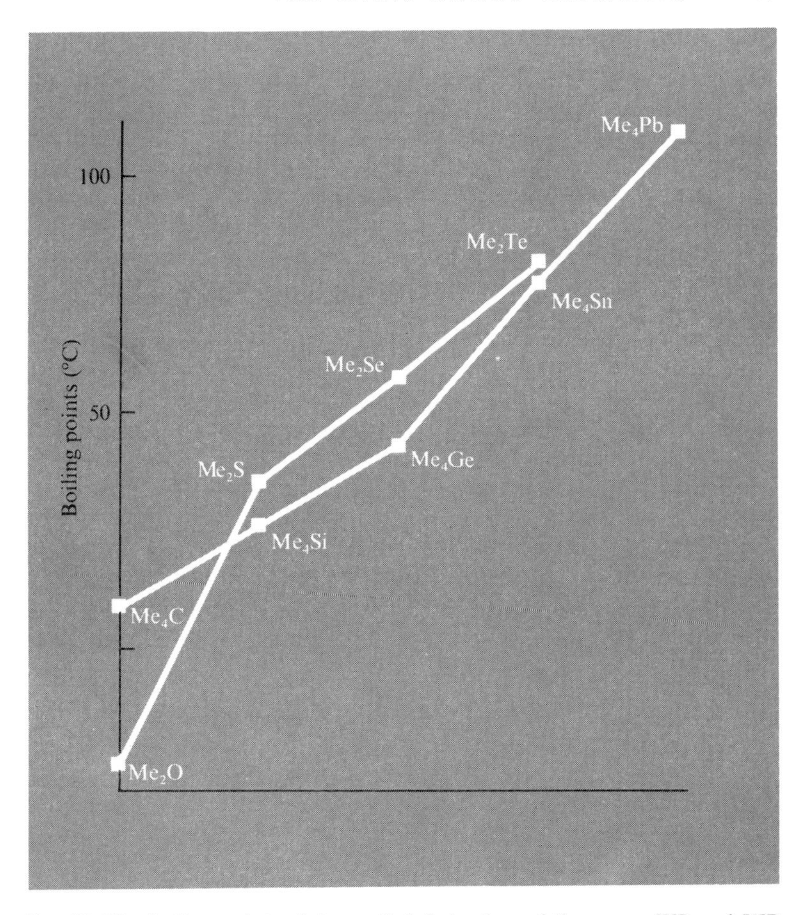

FIG. 21. The boiling points of the methyl derivatives of the group IVB and VIB elements.

forms the basis for organic chemistry. Compounds with linear or cyclic chains of carbon atoms containing single, double, or triple carbon–carbon bonds may be formed as in ethane, $CH_3–CH_3$, cyclopropane, CH_2(—CH_2—CH_2—, ring), ethylene, $CH_2=CH_2$, or acetylene, $CH\equiv CH$.

On descending group IVB, catenated compounds become more difficult to make. Thus silicon forms catenated chlorides up to Si_6Cl_{14}, but germanium forms only Ge_2Cl_6 and tin and lead give no such compounds. Catenation is more frequent in the hydrogen and methyl compounds so that hydrides of germanium are known up to Ge_8H_{18}, and even lead forms the methyl derivative $(CH_3)_6Pb_2$ having a lead–lead bond. This trend is also apparent in group IIIB where only the lightest element, boron, forms an important series of catenated compounds which includes halides such as B_2F_4 and a great number of catenated hydrides. However in other groups this trend may be reversed. Thus in group VB catenated compounds of nitrogen, such as hydrazine, N_2H_4, are no more stable than those of phosphorus (for instance P_2F_4) or arsenic (for instance $As_2(CH_3)_4$), though the catenated compounds of antimony and bismuth are less stable; however in group VIB sulphur, with compounds such as H_2S_2, H_2S_5, and S_2F_{10}, definitely forms more stable catenated compounds than oxygen, whose simplest derivative is hydrogen peroxide, H_2O_2.

Overall the catenated compounds appear to be most common for elements of intermediate electronegativity. Thus carbon forms more compounds than the less electronegative elements of the remainder of group IVB and of groups IA, IIA, and IIIB. In group VIB oxygen is too electronegative and more stable compounds are formed by the less electronegative sulphur and selenium. A simple explanation of this effect is that the most electronegative and least electronegative elements prefer to form ionic or partially ionic bonds, so that they do not give stable catenated compounds which must contain completely non-polar covalent bonds.

Oxides of the elements

Oxygen is the most abundant element on earth and, like fluorine, it forms compounds with almost every other element. Since many of these have quite unusual properties, there is an extensive and varied chemistry of the compounds of oxygen. We shall discuss first the different ways in which oxygen can combine with the elements, and then examine periodic trends in structure and properties of the element oxides.

Ionic oxides

In forming an ionic oxide each oxygen atom gains two electrons, thus giving an O^{2-} ion. As a result of this high negative charge the lattice energies of ionic oxides are high, and ionic oxides are formed from all of the elements which give ionic halides, despite the fact that formation of *isolated* O^{2-} ions from oxygen atoms is an energetically unfavourable process. Another effect of the high lattice energies is to cause the ionic oxides to be rather insoluble in water (since the solvation energy is often lower than the lattice energy), especially if the cation also carries a high charge. In fact the O^{2-} ion is unstable in aqueous solution so that the soluble ionic oxides react with water to form the corresponding hydroxides, according to the equation

$$O^{2-} + H_2O \rightarrow 2OH^-.$$

Covalent oxides

By forming two covalent bonds oxygen attains the noble-gas electron configuration, but this can be achieved in two ways. Thus each oxygen atom can form one double bond as in carbon dioxide, O=C=O, or the oxygen molecule, O=O, or it may form two single bonds as in water, H_2O. Where only element–oxygen double bonds are formed the oxides behave as expected for covalent compounds; they are usually gases, volatile liquids, or low melting-point solids. However when two single element–oxygen bonds are formed, polymeric oxides may result. For example, in silicon dioxide, SiO_2, each oxygen atom is bound to two silicon atoms, each of which in turn is bound to four oxygen atoms, thus giving the polymeric oxide

```
   |       |
 —Si—O—Si—O—
   |       |
   O       O
   |       |
 —Si—O—Si—O—.
   |       |
```

The different structures of CO_2 and SiO_2 result in quite different physical properties for these compounds. Thus CO_2 is a gas which boils at −78°C at atmospheric pressure while SiO_2 is a solid which melts only at 2000°C; this is another example of the first element in a group behaving quite differently from subsequent elements in the group. Because of these great differences in properties it is valuable to classify the covalent oxides as either small covalent oxides such as CO_2 or polymeric covalent oxides such as SiO_2. The full classification of the elements according to the nature of their oxides is shown in Table 5.

TABLE 5

Classification of the element oxides according to structure

Gp IA	Gp IIA	Gp IIB	Gp IIIB	Gp IVB	Gp VB	Gp VIB	Gp VIIB	Gp 0
Li	Be		B	C	N	O	F	
Na	Mg		Al	Si	P	S	Cl	
K	Ca	Zn	Ga	Ge	As	Se	Br	Kr
Rb	Sr	Cd	In	Sn	Sb	Te	I	Xe
Cs	Ba	Hg	Tl	Pb	Bi	Po	At	Rn
			Ionic				*Polymeric covalent*	*Molecular covalent*

TABLE 6

The simple oxides of the elements of groups I–IV

Gp IA	Gp IIA	Gp IIB	Gp IIIB		Gp IVB	
Li_2O	BeO		B_2O_3		CO_2	CO
Na_2O	MgO		Al_2O_3		SiO_2	
K_2O	CaO	ZnO	Ga_2O_3		GeO_2	
Rb_2O	SrO	CdO	In_2O_3	In_2O	SnO_2	SnO
Cs_2O	BaO	HgO	Tl_2O_3	Tl_2O	PbO_2	PbO

Oxides of the group I–IV elements

The oxides of these elements are given in Table 6. With the exception of carbon, all the elements form ionic or polymeric covalent oxides.

The solubility of the oxides in water is strongly dependent on the size and charge of the cation in the ionic oxides, and is greatest for large cations carrying a low charge. Thus on crossing a period the solubilities decrease, so that Cs_2O and BaO are easily soluble in water (with reaction, p. 48 while HgO, Tl_2O_3, and PbO_2 are insoluble. Similarly, on descending a group the solubilities increase, so that MgO is insoluble in water while BaO dissolves readily.

The heavier elements of groups IIIB and IVB form oxides both in the group oxidation state (for instance $Sn(IV)O_2$) and in the oxidation state two less than the group number (for instance Sn(II)O). The relative stability of the lower and higher oxidation state oxides is similar to that of the corresponding element fluorides (p. 39). Carbon, the first element in group IV, differs completely from the other elements in these groups in forming the molecular covalent oxides carbon dioxide, CO_2, and carbon monoxide, CO.

Oxides of the group V elements

All of the elements of this group form oxides of the type E_2O_3 (E = group V element) and all but bismuth also form the oxides E_2O_5. However the structures and properties of these oxides are often quite different.

Nitrogen also forms a unique series of oxides which includes dinitrogen monoxide, N_2O, nitrogen oxide, NO, nitrogen dioxide, NO_2, and dinitrogen tetroxide, N_2O_4, all of which are small covalent molecules in the gas phase. Some of these oxides contain an 'odd' or unpaired electron. Thus we may write the electronic structure of nitrogen dioxide as

$$\begin{array}{l} \cdot\underset{\cdot\cdot}{N}::\ddot{O}: \\ :\underset{\cdot\cdot}{O}: \end{array} \quad \text{or} \quad \begin{array}{l} \cdot N{=}O \\ \;\downarrow \\ \;O \end{array}$$

with an odd electron on nitrogen. The nitrogen atom can gain the stable octet of electrons if nitrogen dioxide dimerizes, and indeed nitrogen dioxide exists in equilibrium with its dimer, dinitrogen tetroxide:

$$O_2N\cdot + \cdot NO_2 \rightleftharpoons O_2N{-}NO_2.$$

The oxides of phosphorus, which are formed by burning the element in air, are best written as P_4O_6 and P_4O_{10}, since the solid compounds contain the molecular units:

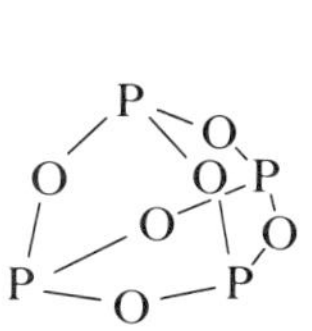

and

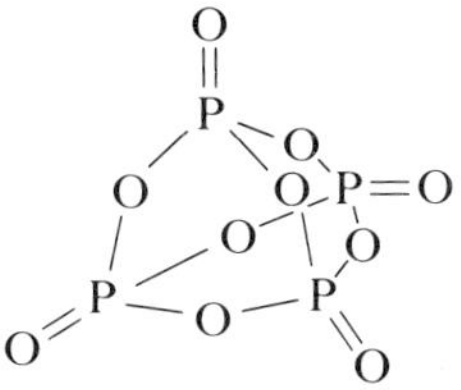

It can be seen that the tetrahedral P_4 units present in yellow phosphorus are maintained, with oxygen atoms bridging between pairs of phosphorus atoms. In P_4O_{10} each phosphorus is bound to an additional 'terminal' oxygen atom.

Thus the oxides of phosphorus are on the borderline between the polymeric covalent oxides, with infinite networks of atoms, and the small, covalent oxides. The oxides of arsenic and antimony are similar, but the only stable oxide of bismuth, Bi_2O_3, is an ionic compound.

Oxides of the group VIB, VIIB, and 0 elements

The well-characterized oxides of these elements are listed in Table 7.

The group VIB elements form an important group of oxides. Oxygen, like fluorine, always causes an expansion of the valency shell of the heavier group VIB elements beyond the octet, so that sulphur forms the oxides sulphur dioxide, SO_2, and sulphur trioxide, SO_3, but *not* sulphur monoxide,

TABLE 7

Oxides of the elements of groups VIB, VIIB, and 0

Gp VIB		Gp VIIB					Gp 0
O_2		F_2O					
SO_2	SO_3	Cl_2O	ClO_2		Cl_2O_6	Cl_2O_7	
SeO_2	SeO_3	Br_2O	BrO_2				
TeO_2	TeO_3		I_2O_4	I_2O_5			XeO_3
PoO_2							

S=O, which would be a sulphur analogue of the oxygen molecule. On descending the group the following trends may be seen:

(1) The stability of element–oxygen double bonds decreases so that sulphur dioxide and sulphur trioxide contain S=O double bonds and are volatile (the vapours contain discrete SO_2 or SO_3 molecules), while SeO_2 and TeO_2 are solid polymeric oxides.

(2) The stability of the oxidation state VI decreases, so that SeO_3 and TeO_3 are much less stable than SO_3.

The group VIIB elements form a wide range of oxides, as shown in Table 7, but all are unstable compounds and many are explosive. No meaningful periodic trends can be found. The only important noble-gas oxide is XeO_3 which is an explosive solid formed by hydrolysis of XeF_6.

Acidic and basic oxides

The classification of the oxides of the elements as acidic or basic is dependent on their reactions with water. It is well-known that water undergoes a self-ionization process:

$$2H_2O \rightarrow H_3O^+ + OH^-.$$

In water the product $[H_3O^+][OH^-]$ is a constant† and in pure water $[H_3O^+] = [OH^-] = 10^{-7}$ M. If addition of another substance to water causes an increase in $[H_3O^+]$ that substance is an acid while if it causes an increase in $[OH^-]$ it is a base.

Thus sodium oxide is a basic oxide since it dissolves in water according to the equation

$$Na_2O + H_2O \rightarrow 2Na^+ + 2OH^-,$$

and causes an increase in hydroxide ion concentration.

On the other hand, when sulphur trioxide, SO_3, reacts with water there is no tendency for a similar ionization to occur giving the S^{6+} ion. Rather, the sulphur atom, having a high polarizing power, accepts a further electron pair from an oxygen atom of a water molecule into a vacant *d* orbital and gives sulphuric acid according to the equation

$$SO_3 + 3H_2O \rightarrow SO_4^{2-} + 2H_3O^+.$$

Thus the oxides Na_2O and SO_3 show extreme differences in their reactions with water, the former giving a strongly basic solution and the latter a strongly acidic solution. Basic oxides and acidic oxides react with one another to give salts, for example

$$Na_2O + SO_3 \rightarrow Na_2SO_4.$$

Certain oxides react with both basic and acidic oxides in solution. For instance, although aluminium oxide is itself too insoluble to react readily with

† The brackets represent concentrations. Thus $[H_3O^+]$ is a shorthand way of writing 'concentration of H_3O^+ ions.'

either acids or bases, aluminium hydroxide, $Al(OH)_3$, dissolves both in sulphuric acid to give aluminium sulphate (remember that the ions H^+ and Al^{3+} are in fact strongly hydrated, p. 43),

$$Al(OH)_3 + 3H^+ \rightarrow Al^{3+} + 3H_2O,$$

and in sodium hydroxide to give sodium aluminate,

$$Al(OH)_3 + 3OH^- \rightarrow Al(OH)_6^{3-}.$$

The former is a reaction typical of basic oxides or hydroxides while the latter is typical of acidic oxides. Oxides like aluminium oxide, which show both acidic and basic properties, are called *amphoteric* oxides.

We may therefore further classify the elements according to whether their oxides are basic, acidic, or amphoteric. This classification is given in Table 8, and the similarities with the classification in terms of structure and bonding in Table 5 (p. 49) are obvious. Thus the ionic oxides are usually basic though, if the central metal carries a high charge as in group III, they may be amphoteric. The polymeric covalent oxides show either amphoteric or acidic properties, while the molecular covalent oxides are almost always acidic.

Like the parent acidic oxides, the oxyanions such as the sulphate ion, SO_4^{2-} (from SO_3), or the carbonate ion, CO_3^{2-} (from CO_2), may exist as molecular ions with multiple element–oxygen bonds or as polymeric ions with bridging oxygen atoms. Thus the molecular covalent oxide CO_2 gives the molecular carbonate ion CO_3^{2-}, while the polymeric oxide SiO_2 always gives polymeric silicates, which, like silicon dioxide itself, are highly insoluble and are often present in rocks.

Sulphides of the elements

The element sulphides can be classified, like the oxides, as ionic, molecular covalent, or polymeric covalent sulphides. However, for a number of reasons,

TABLE 8

Classification of the element oxides

Gp IA	Gp IIA	Gp IIB	Gp IIIB	Gp IVB	Gp VB	Gp VIB	Gp VIIB
Li	Be		B	C	N	O	F
Na	Mg		Al	Si	P	S	Cl
K	Ca	Zn	Ga	Ge	As	Se	Br
Rb	Sr	Cd	In	Sn	Sb	Te	I
Cs	Ba	Hg	Tl	Pb	Bi	Po	At
	Basic oxides			*Amphoteric oxides*		*Acidic oxides*	

the periodic distribution of these structural types is quite different from that for the oxides.

Ionic sulphides

The only truly ionic sulphides are those of the group IA and group IIA elements. Thus there are far fewer ionic sulphides than there are ionic oxides. Several factors may account for this difference.

(1) Since the S^{2-} ion is larger than the O^{2-} ion, lattice energies for ionic sulphides will be lower than for oxides, so that they are less likely to be formed.

(2) The larger S^{2-} ion is much more polarizable than the O^{2-} ion so that an increased tendency to form covalent compounds is expected. This trend is also anticipated from the fact that sulphur is less electronegative than oxygen.

Covalent sulphides

Sulphur forms stable double bonds only to the elements oxygen and carbon. Thus the only simple molecular sulphide is formed by carbon which gives carbon disulphide, CS_2, a volatile liquid which has the structure S=C=S.

Since there are fewer ionic sulphides and molecular covalent sulphides in comparison with the oxides, it follows that there are more polymeric covalent sulphides. However these polymeric sulphides exist in a wide variety of structures, so that it is difficult to find meaningful periodic trends and we shall not discuss them further.

The diagonal relationship

We have already seen the value of the concept of the polarizing power of a cation in interpreting the chemical and physical properties of chemical compounds. Thus we have discussed covalency in metal halides, hydrides, oxides, and sulphides and the reactions of metal oxides and halides with water in terms of the polarizing power of the cation. In general, we expect compounds of cations with similar polarizing power to have similar properties.

Now we have seen (p. 21) that the polarizing power of a cation can be measured approximately by the ratio of its charge to its radius, and it can readily be calculated that elements on diagonal lines in the periodic table, as shown below, will give cations having similar polarizing powers.

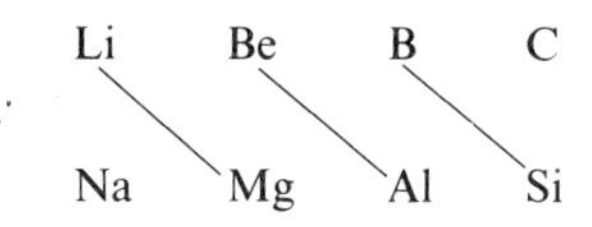

Thus the ion Al^{3+} is larger but carries a higher charge than the ion Be^{2+}, so that these ions have similar polarizing powers, and we expect similarities in the chemistry of these elements. This is indeed the case and, in fact, beryllium resembles aluminium more closely than it resembles magnesium or the other group II elements. Both beryllium and aluminium are metallic elements; they are not attacked by air or water, owing to the formation of an inert oxide film on the metallic surface; both react with strong bases to give respectively beryllates and aluminates. Both elements form hard, unreactive, very high melting-point oxides and amphoteric hydroxides which are insoluble in water. The anhydrous halides of both elements are covalent with bridging halogen atoms; solutions of the halides in water are acidic and the hydrated salts are obtained on crystallization. The chief difference between the elements is that aluminium can form compounds with co-ordination number six while the maximum co-ordination number for beryllium, as for other elements of the first short period, is four. Thus in AlF_3 each Al atom is six co-ordinate while in BeF_2 the Be atoms are four co-ordinate. On crystallization from water aluminium salts usually contain the hydrated cation $Al(OH_2)_6^{3+}$ while beryllium gives only the $Be(OH_2)_4^{2+}$ ion.

Lithium resembles magnesium less strongly, and the differences can be attributed to the greater polarizing power of Mg^{2+} compared with Li^+. Thus lithium salts are more typically ionic than magnesium salts; in aqueous solution the Li^+ ion is solvated less strongly than Mg^{2+}, and lithium hydroxide is a stronger base than magnesium hydroxide.

There are strong similarities between the chemistries of boron and silicon. Thus both elements are metalloids; they give covalent monomeric chlorides which are completely hydrolysed by water to give boric acid or silicic acid and hydrogen chloride, and both form acidic oxides.

More generally, we have seen that the metalloid elements occur on a diagonal line across the periodic table (Table 1, p. 28) as do the polymeric covalent oxides (Table 5, p. 49) or amphoteric oxides (Table 8, p. 53), as well as the hydrides with intermediate properties between the typical ionic and covalent hydrides (Table 4, p. 44).

This diagonal relationship is most useful for the elements of groups I–IV, and is much less important for the most electronegative elements, which have little or no tendency to form cations. Thus there is little resemblance between the compounds of carbon and phosphorus or between those of nitrogen and sulphur.

PROBLEMS

6.1. Explain why in covalent compounds bromides usually have higher boiling points than chlorides, while in ionic compounds the reverse is true.

6.2. Which of the following halides do you expect to be hydrolysed rapidly by water?

$$BCl_3, CCl_4, SiCl_4, PCl_5, SF_6$$

6.3. An element is a metalloid. What properties do you expect its halides, hydride, and oxide to possess? (Hint: compare Tables 1, 2, 4, 5, and 8.)

7. Periodicity and Chemical Compounds: The Transition Elements

Introduction

MANY compounds of the transition elements have certain properties which are not possessed by compounds of the main-group elements, owing to the presence of partially-filled shells of *d* electrons on the transition-element ions.

However there are also similarities between the transition elements and the main-group elements, and, before treating the compounds of the transition elements in more detail, it will be valuable to make a brief general survey of the similarities and differences between the chemistries of the elements. Since the transition elements are all metallic in nature, the most useful comparisons can be made with the metallic main-group elements of groups I–IV.

The consecutive ionization energies of the elements of the first long period from potassium to germanium are shown in Fig. 22. There is usually an obvious 'break' in the separations between consecutive ionization energies for a main group element. Thus for potassium there is an obvious break between the first and second ionization energies, so that K^+ is easily formed but K^{2+} is never present in chemical compounds. Similarly the calcium atom easily loses two electrons to reach the (argon) noble-gas electron configuration, but a prohibitive amount of energy would be required to form the Ca^{3+} ion. For the group IIIB element gallium this break, which is less marked than for the group IA or IIA elements, comes between the third and fourth ionization energy so that gallium prefers to adopt the group oxidation state of three.

These trends are still present in the early part of the transition series. For the group IIIA element scandium there is an obvious break between the third and fourth ionization energies and, as we might expect, the chemistry of scandium is dominated by the tendency to form the Sc^{3+} ion. This ion has *no d* electrons so that compounds of scandium(III) show few typical 'transition' properties.

On passing across the transition series the separations between subsequent ionization energies increase and it becomes progressively more difficult to form transition-metal ions in their group oxidation states. Therefore compounds of titanium(IV) are more stable than those of vanadium(V), which in turn are more stable than those of chromium(VI) or manganese(VII), while no compounds of iron(VIII) are known. It follows that the formation of stable compounds in oxidation states lower than the group number becomes increasingly important after scandium.

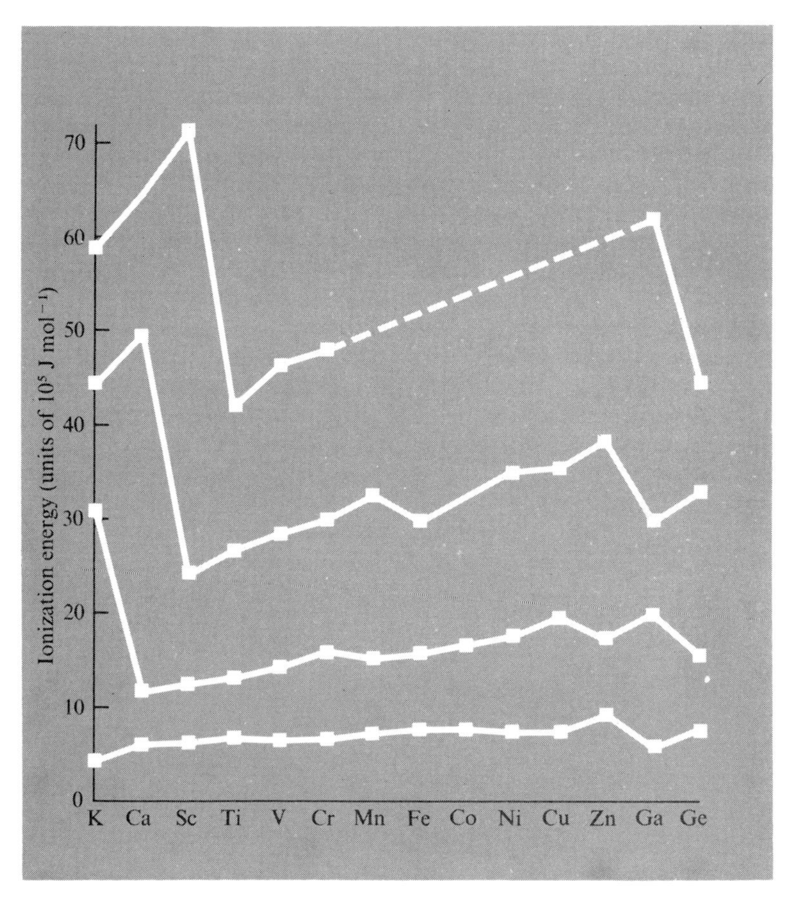

FIG. 22. Ionization energies of the elements of the first long period. Consecutive ionization energies increase so that the first ionization energies lie on the bottom line and the fourth ionization energies (where known) on the top line.

Thus titanium forms not only the chloride $TiCl_4$ but also $TiCl_3$ and $TiCl_2$. In $TiCl_4$ the electron configuration at titanium(IV) is (Ar core) and there are no *d* electrons. Titanium tetrachloride has many similarities with the group IVB tetrachlorides such as $SnCl_4$; both are colourless liquids which are violently hydrolysed by water. However, in $TiCl_3$ or $TiCl_2$ the electron configuration at titanium is respectively (Ar core) $3d^1$ and (Ar core) $3d^2$. It is the presence of these odd *d* electrons which leads to the characteristic colours and magnetic properties of the transition-element compounds. Thus

$TiCl_3$ is a violet solid and $TiCl_2$ is a black solid; both are quite different from $SnCl_2$, which is white.

The typical 'transition' property of variable oxidation state then arises for two reasons: first because it is often difficult to reach the group oxidation state, and secondly because the differences in stability between the various lower oxidation-state compounds are small. The stable oxidation states often vary in units of one, rather than in units of two as found for the main-group elements. Thus, as noted above, titanium often forms compounds in oxidation states IV, III, and II, while tin forms compounds in oxidation states IV and II only.

On crossing the transition series each successive element can form compounds in a wider range of oxidation states. Thus manganese compounds in all oxidation states from −I to VII are known, but after manganese the elements iron, cobalt, and nickel never form compounds in the group oxidation state, so that for these elements some of the *d* electrons never take part in bonding and are effectively core electrons. In fact on crossing the transition series it seems that the 3*d* orbitals are stabilized with respect to the 4*s* and 4*p* orbitals, as shown diagrammatically in Fig. 23, and by the end of the transition series they have entered the electron core.

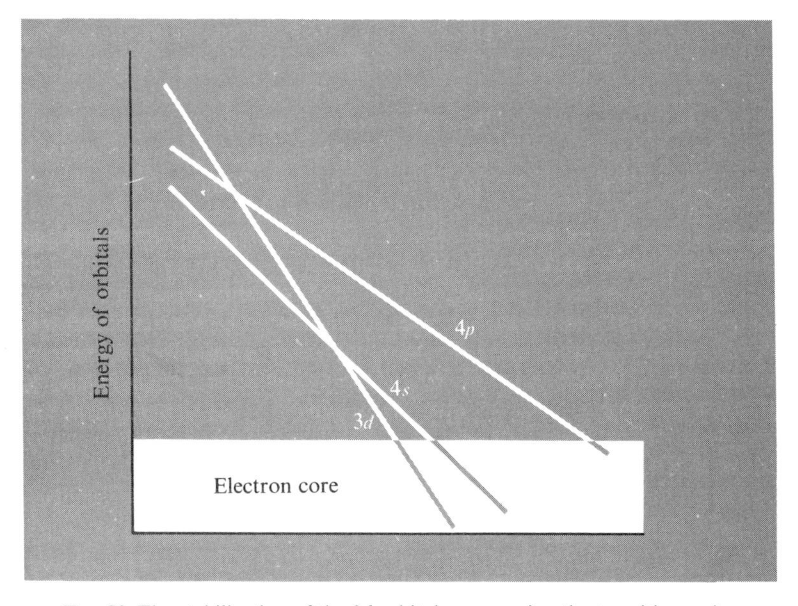

FIG. 23. The stabilization of the 3*d* orbitals on crossing the transition series.

The elements of group IIB at the end of the transition series very rarely use their d orbitals in bonding and, as we have seen (p. 34), they show no transition-metal properties. Overall then, the first elements and the last elements of the transition series behave like main-group metallic elements, but the inner transition elements, which commonly form ions having partially-filled shells of d electrons, have the extra property of forming compounds, which are often highly coloured, in a wide range of oxidation states.

Binding energies of the transition elements

The elemental binding energies (p. 26) of the transition elements again show evidence that the $3d$ electrons enter the electron core towards the end of the transition series. Binding energies for the first and third transition series are given in Fig. 24. It can be seen that at the beginning of the transition series the binding energies increase with the number of electrons, indicating that all the d electrons contribute to the binding. However towards the end of each series there is a drop in binding energy as progressively more d electrons enter the electron core and no longer contribute to the binding. At the end of the transition series (Group IIB) only the $4s$ electrons contribute to the binding energy, which is so low at the end of the third transition series that mercury is a liquid at room temperature. This may be described as an

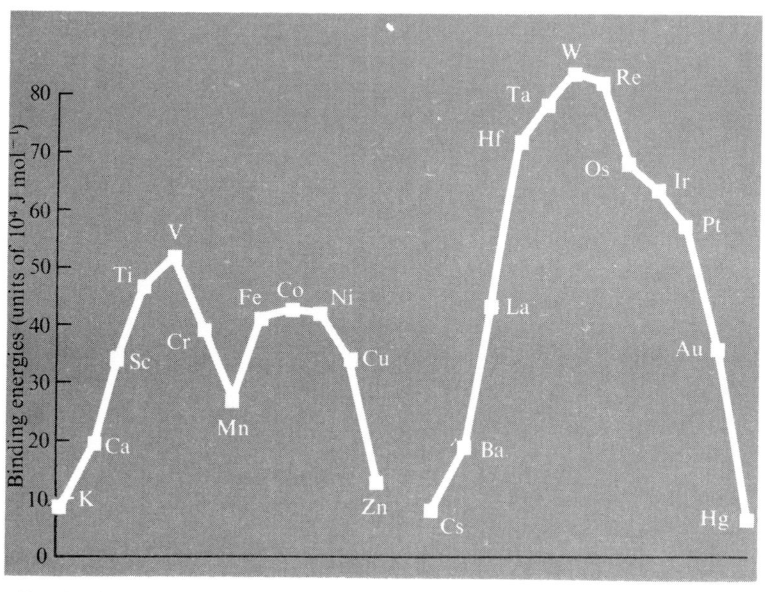

FIG. 24. Binding energies for the elements of the first and third transition series.

example of the inert-pair effect (p. 39). It is because the 3*d* electrons have entered the electron core that the group IIB elements show no typical transition-metal properties.

We can gain our first insight into another general trend in transition-metal chemistry by comparison of the binding energies for the first and third transition series. In the first transition series the general increase in binding energies ends at vanadium, and vanadium is the last element to form stable compounds in the group oxidation state. Vanadium(V) compounds are quite stable, while subsequent elements in their group oxidation states (for instance Cr(VI)) are strongly oxidizing, as they are easily reduced to lower oxidation states. In the third transition series the general increase in binding energies on crossing the periodic table ends at group VIA, and it seems that for tungsten all the 5*d* electrons contribute to the binding. All the 5*d* electrons can also be used in forming compounds with other elements, so that tungsten(VI) compounds are quite stable compared with those of chromium(VI).

This increased stability of high oxidation states for the elements of the second and third transition series compared with the first is an important feature of transition-metal chemistry.

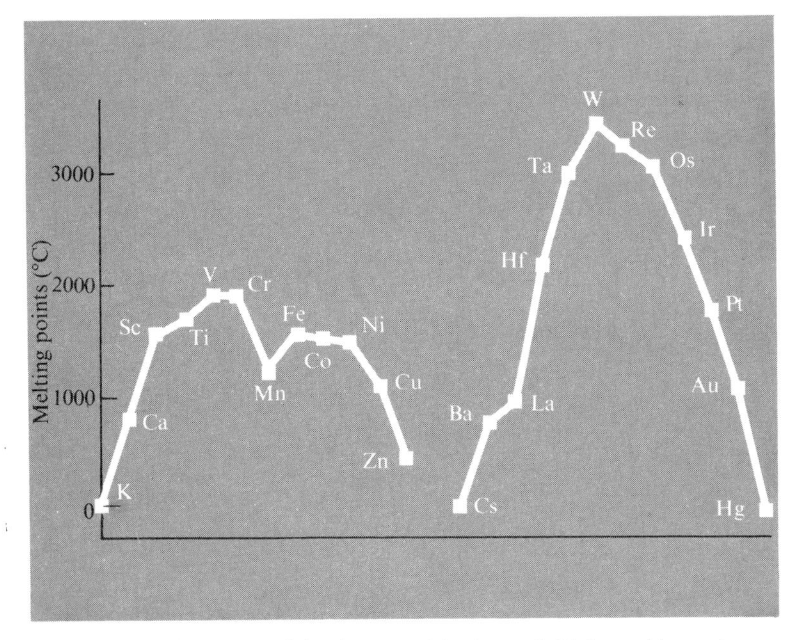

FIG. 25. Melting points of the elements of the first and third transition series.

Finally in this section, we note that the binding energies can be correlated with many of the physical properties of the transition elements. The melting points of the first and third transition series, shown in Fig. 25, correlate very well with the binding energies in Fig. 24. We saw similar correlations for the main group elements.

Oxidation states of the transition elements

We have already predicted some of the trends in oxidation states of the transition elements. Let us now examine these trends more closely. In Table 9 are listed those chlorides of the transition metals which are sufficiently stable to be isolated. The property of multiple oxidation states of the transition metals is well illustrated by the great number of compounds formed. We also see that higher oxidation states are favoured by the second and third transition-series elements.

Another feature is the evident high stability of the oxidation state II for the chlorides of the first transition series. We might have predicted this from the ionization energies in Fig. 22 (p. 57). The third ionization energies for the first transition-series elements are much higher than the first two ionization energies. To form the dipositive ions the two 4s electrons are generally lost and the ionization process is

$$[\text{Ar core}]\, 3d^n 4s^2 \rightarrow [\text{Ar core}]\, 3d^n.$$

The 4s electrons are removed *before* the 3d because the latter are strongly stabilized as the effective nuclear charge increases. This stabilization also means that a high energy is required to remove a further electron to form the tripositive ion.

In the second and third transition series the stabilities of the 4d and 5d orbitals are not so dependent on the nuclear charge, so that third and subsequent ionization potentials are lower than for the first transition series; this can be seen by comparing the ionization energies for the second transition series in Fig. 26 with those for the first transition series in Fig. 22 (p. 57). For this reason the oxidation state II is not so common for the heavier transition elements.

High oxidation states

As is the case with the main group elements, the highest oxidation states are attained when the transition metals combine with fluorine or oxygen. The highest oxidation-state fluorides of the transition elements are shown in Table 10. Group oxidation states are commonly found up to group VIIA in which rhenium forms ReF_7. Thereafter high oxidation states become progressively less stable as the d orbitals enter the electron core.

In combination with oxygen yet higher oxidation states may be attained. Thus in the first transition series the group oxidation state is achieved by

TABLE 9

Chlorides of the transition elements

Gp IIIA	Gp IVA	Gp VA	Gp VIA	Gp VIIA	Gp VIII	Gp VIII	Gp VIII	Gp IB	Gp IIB
$ScCl_3$	$TiCl_2$ $TiCl_3$ $TiCl_4$	VCl_2 VCl_3 VCl_4	$CrCl_2$ $CrCl_3$	$MnCl_2$	$FeCl_2$ $FeCl_3$	$CoCl_2$ $CoCl_3$	$NiCl_2$	$CuCl$ $CuCl_2$	$ZnCl_2$
YCl_3	$ZrCl_2$ $ZrCl_3$ $ZrCl_4$	$NbCl_3$ $NbCl_4$ $NbCl_5$	$MoCl_2$ $MoCl_3$ $MoCl_4$ $MoCl_5$	$TcCl_4$ $TcCl_6$	$RuCl_3$ $RuCl_4$	$RhCl_3$	$PdCl_2$	$AgCl$	$CdCl_2$
$LaCl_3$	$HfCl_3$ $HfCl_4$	$TaCl_2$ $TaCl_3$ $TaCl_4$ $TaCl_5$	WCl_2 WCl_4 WCl_5 WCl_6	$ReCl_3$ $ReCl_5$	$OsCl_2$ $OsCl_3$ $OsCl_4$	$IrCl_3$	$PtCl_2$ $PtCl_4$	$AuCl$ $AuCl_3$	Hg_2Cl_2 $HgCl_2$

TABLE 10

Highest oxidation-state fluorides of the transition elements

Gp IIIA	Gp IVA	Gp VA	Gp VIA	Gp VIIA	Gp VIII	Gp VIII	Gp VIII	Gp IB	Gp IIB
ScF_3	TiF_4	VF_5	CrF_3	MnF_3	FeF_3	CoF_3	NiF_2	CuF_2	ZnF_2
YF_3	ZrF_4	NbF_5	MoF_6	TcF_6	RuF_6	RhF_6	PdF_3	AgF_2	CdF_2
LaF_3	HfF_4	TaF_5	WF_6	ReF_7	OsF_6	IrF_6	PtF_6	AuF_3	HgF_2

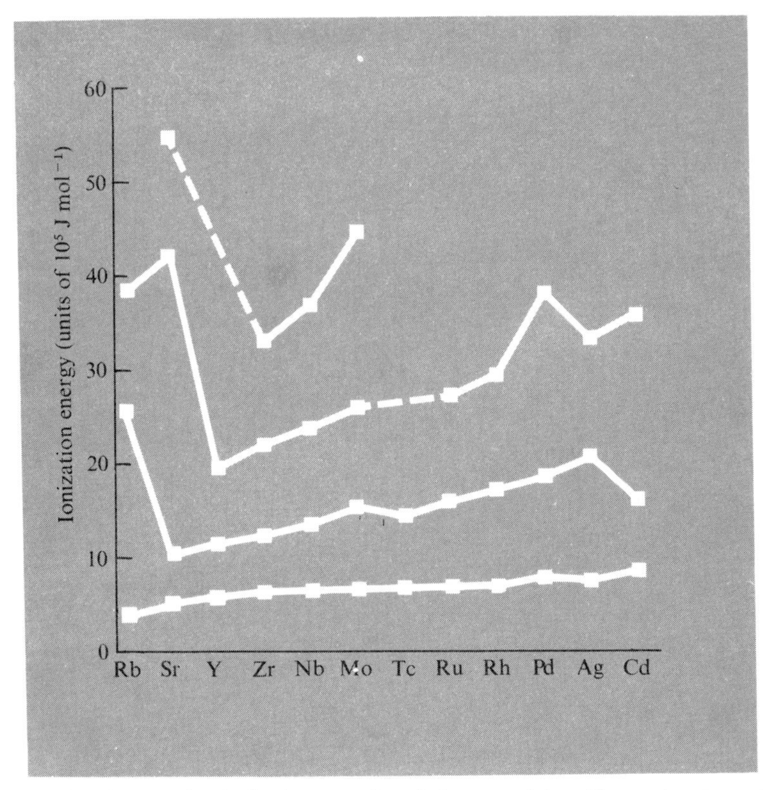

FIG. 26. Consecutive ionization energies of the second transition-series elements (compare with Fig. 22, p. 57).

chromium in chromium trioxide, CrO_3, and the chromate ion, CrO_4^{2-}, and by manganese in the permanganate ion, MnO_4^-, which as potassium permanganate is an important oxidizing agent. In the third transition series the highest group oxidation state is found for ruthenium and osmium which form the tetroxides RuO_4 and OsO_4, derived from Ru(VIII) and Os(VIII) respectively.

Ionic and covalent bonding

The nature of the bonding in the transition-metal halides and oxides is determined by the same principles which we saw governing that in the corresponding compounds of the main-group elements. For instance we expect and find a greater tendency to covalency when the cation carries a higher charge. In agreement with this it is found that the titanium chlorides $TiCl_2$

and $TiCl_3$ are ionic compounds with high melting points, while $TiCl_4$ is a volatile liquid with covalent titanium–chlorine bonds. There are many similar examples with other transition-metal halides.

The properties of the transition-metal oxides again depend strongly on the oxidation state of the metal. The oxides in low oxidation states are usually rather unstable, but are typically basic. For instance nickel(II) oxide dissolves readily in acids to give other nickel(II) salts. Oxides in intermediate oxidation states are often amphoteric, a typical example being chromium(III) oxide, Cr_2O_3, which gives chromium(III) salts with acids and chromites with bases. Finally in high oxidation states the oxides are acidic. For instance chromium(VI) oxide, CrO_3, dissolves in basic solution to give the chromate ion, CrO_4^{2-}, and manganese(VII) oxide, Mn_2O_7, readily gives the permanganate ion, MnO_4^-. In this sense there is some formal resemblance to the main-group analogues of these compounds. Thus in group VIB sulphur trioxide, SO_3, gives the sulphate ion, SO_4^{2-}, and in group VIIB chlorine heptoxide, Cl_2O_7, gives the perchlorate ion, ClO_4^-. However there are few chemical similarities between sulphates and chromates, or between perchlorates and permanganates.

Covalent and ionic radii

The covalent radii and ionic radii (for the dipositive cations) of the elements of the first transition series are given in Fig. 27. It can be seen that the covalent radii decrease rapidly at the start of the series, then become substantially constant, and finally begin to increase at the end of the series when the filled 3*d* orbitals have contracted into the electron core and so shield the outer 4*s* electrons more effectively from the nucleus. Changes in the ionic radii on crossing the series are much less regular, so that periodic trends in the properties of these ions are difficult to rationalize. There is, however, a steady decrease in ionic radius between Mn^{2+} and Ni^{2+} and therefore a corresponding increase in the polarizing power of these ions. In turn there is a steady increase in the importance of covalency between the compounds of manganese(II) and nickel(II), and the stability of the hydrated ion $M(OH_2)_6^{2+}$ follows the series $Mn^{2+} < Fe^{2+} < Co^{2+} < Ni^{2+} < Cu^{2+} > Zn^{2+}$ in accordance (with the exception of Cu^{2+}) with the predicted polarizing powers of these ions.

Another useful comparison can be made between the covalent radii of the first and subsequent transition-series elements. These are given in Table 11.

It can be seen that there is the expected increase in radius on passing from the first to the second transition series, but that, within a group, the values for the second and third transition-series elements are almost the same. The reason for this is that the lanthanide elements, in which the 4*f* electron shell is filled, precede the elements of the third transition series. These 4*f* electrons shield outer electrons inefficiently and the resulting increased effective nuclear

TABLE 11

Covalent radii of the transition elements (units of nm)

Gp IIA	Gp IIIA	Gp IVA	Gp VA	Gp VIA	Gp VIIA	Gp VIII	Gp VIII	Gp VIII	Gp IB	Gp IIB
Ca 0·174	Sc 0·144	Ti 0·132	V 0·122	Cr 0·118	Mn 0·117	Fe 0·117	Co 0·116	Ni 0·115	Cu 0·117	Zn 0·125
Sr 0·191	Y 0·162	Zr 0·145	Nb 0·134	Mo 0·130	Tc 0·127	Ru 0·125	Rh 0·125	Pd 0·128	Ag 0·134	Cd 0·148
Ba 0·198	La 0·169	Hf 0·144	Ta 0·134	W 0·130	Re 0·128	Os 0·126	Ir 0·127	Pt 0·130	Au 0·134	Hg 0·149

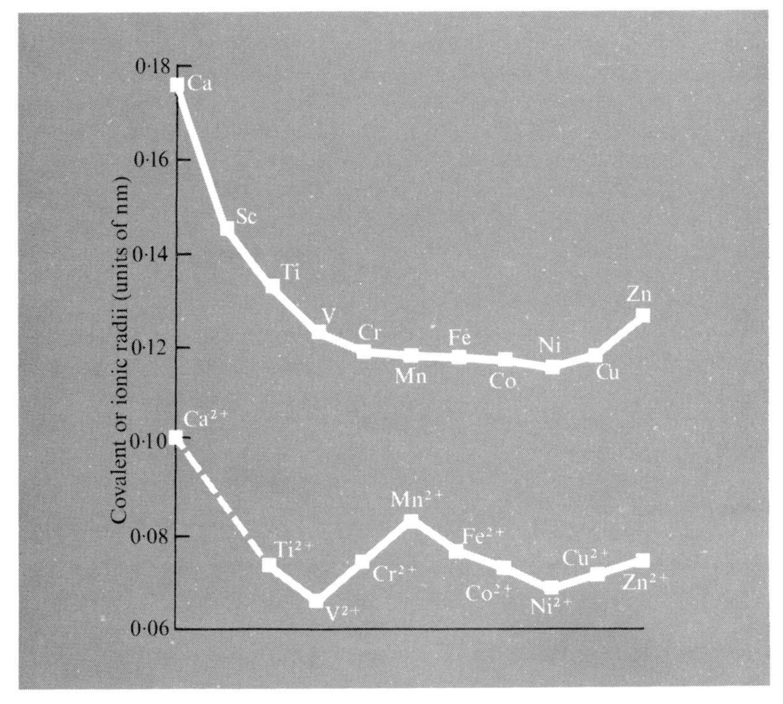

FIG. 27. Covalent radii and ionic radii (for M^{2+}) of first transition-series elements.

charge causes a general contraction of the orbitals, so that the radii of the elements of the third transition series are somewhat lower than would be expected by analogy with the behaviour of main-group elements. The contraction is such that the covalent radii (and ionic radii) for the third transition series are very close to those for the second transition series. Although the elements of the second and third transition series give larger ions, covalency is more important for them than for the elements of the first transition series. For instance while nickel(II) chloride, $NiCl_2$, is an essentially ionic compound, palladium(II) and platinum(II) chlorides exist as covalent, chlorine-bridged polymers.

The reasons for this effect are rather complex. One reason is that the first ionization energies and the electron affinities for the second and third transition series are often higher than for the first series so that, in contrast to the usual main-group trends, the larger element atoms are the more electronegative and so form predominantly covalent compounds. In the cations of the first transition series the $3d$ orbitals contract into the electron core to a greater extent and are therefore less readily available to form covalent bonds than the 4*d* or 5*d* orbitals of the second and third transition series.

'Class *a*' and 'Class *b*' ions

Some of the heavier transition-metal cations have a greater affinity for large anions (such as S^{2-} or I^-) than for small anions (such as O^{2-} or F^-) in aqueous solution. This is partly attributable to the larger orbitals of these cations which, in forming covalent bonds, give better overlap if the orbitals on the anion are of similar size, i.e. if the anion is also large. These metal ions are called 'Class *b*' metal ions, while metal ions which prefer small anions are called 'Class *a*' metal ions. The full classification is shown in Table 12.

Since the Class *b* metal ions have little affinity for oxygen donors (such as H_2O) their halides and sulphides are insoluble in water. For instance nickel(II) chloride (Class *a*) dissolves readily in water giving the $Ni(OH_2)_6^{2+}$ ion, but palladium(II) and platinum(II) chlorides (Class *b*) are insoluble. Advantage may be taken of this property in quantitative or qualitative analysis. Thus the insolubility of the silver halides is used in quantitative analysis for silver or for halide ions, while the insolubility of mercury(II) or lead(II) sulphide forms the basis for qualitative tests for these metals.

Transition-metal complexes

Before discussing the colours and magnetic properties of transition-metal compounds, it will be necessary to survey an important class of compounds (known as *complexes*) formed by the transition elements.

Complexes of main group elements are usually called addition compounds. Thus boron trifluoride adds ammonia according to the equation

$$H_3N:+BF_3 \rightarrow H_3N\rightarrow BF_3.$$

The B–N bond is an electron-pair bond in which both electrons are supplied by the nitrogen atom; such a bond is commonly called a *co-ordinate bond* and the product may be called a *co-ordination complex.*

The boron trifluoride ammonia complex is neutral, but positively or negatively charged *complex ions* may also be formed. Thus the ions $SnCl_6^{2-}$ or $Al(OH_2)_6^{3+}$ may be considered to be complex ions.

The transition elements form an especially important variety of complex compounds and complex ions, in which the co-ordination number six is very common. For instance chromium(III) chloride dissolves in water to give the complex ion $Cr(OH_2)_6^{3+}$, from which the water molecules can be displaced by ammonia according to the equation

$$Cr(OH_2)_6^{3+} + 6NH_3 \rightarrow Cr(NH_3)_6^{3+} + 6H_2O.$$

In transition-metal complexes the groups attached to the metal are called *ligands* so that the ions $Cr(OH_2)_6^{3+}$ or $Cr(NH_3)_6^{3+}$ contain the ligands water

TABLE 12

The 'Class a' and 'Class b' metal ions. Certain elements on the borderline, for example, Tc and Re, may show either 'Class a' or 'Class b' properties

IA	IIA	IIIA	IVA	VA	VIA	VIIA	VIII	VIII	VIII	IB	IIB	IIIB	IVB	VB
Li	Be											B	C	N
Na	Mg											Al	Si	P
K	Ca	Sc	Ti	V	Cr	Mn	Fe	Co	Ni	Cu	Zn	Ga	Ge	As
Rb	Sr	Y	Zr	Nb	Mo	Tc	Ru	Rh	Pd	Ag	Cd	In	Sn	Sb
Cs	Ba	La	Hf	Ta	W	Re	Os	Ir	Pt	Au	Hg	Tl	Pb	Bi
			Class a								*Class b*			

or ammonia respectively. Ligands can be either neutral or negatively charged; thus the complex ion $PtCl_6^{2-}$ would be derived from the Pt^{4+} cation surrounded by six Cl^- ligands.

In these six co-ordinate complexes the ligands are almost always at the corners of an octahedron with the metal atom at the centre. The $Cr(NH_3)_6^{3+}$ ion is represented below:

NH_3
H_3N NH_3
Cr
H_3N NH_3
NH_3
$3+$

If the attraction between the metal ion and the ligands is considered to be electrostatic in origin, the colours and magnetic properties of transition-metal compounds can be explained in terms of a simple bonding theory known as crystal-field theory.

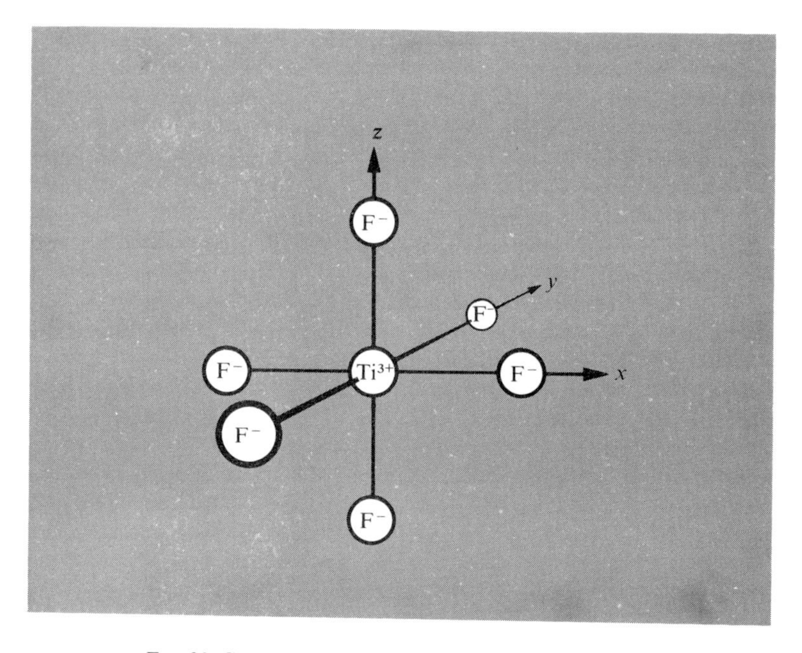

FIG. 28. Crystal-field representation of the TiF_6^{3-} ion.

Crystal-field theory

Let us consider the complex ion TiF_6^{3-} to be formed from the cation Ti^{3+}, whose electron configuration is (Ar core) $3d^1$, by bringing up six F^- ligands along the x, y, and z axes as shown in Fig. 28.

In the free ion Ti^{3+} the odd d electron has an equal chance of being in any of the five $3d$ orbitals since they are of equal energy. However an examination of the representations of d orbitals in Fig. 4 shows that in the complex TiF_6^{3-} the lobes of the $d_{x^2-y^2}$ and d_{z^2} orbitals are directed *along* the axes and so directly *towards* the F^- ions while those of the d_{xy}, d_{yz}, and d_{xz} orbitals are directed *between* the axes and so between the F^- ions. The electron will prefer to occupy an orbital whose lobes project between the F^- ions since the repulsive electrostatic forces between the electron and the negatively charged ligands will then be minimized. Thus the d orbitals are split into two sets of unequal energy by the *octahedral crystal field* as shown below:

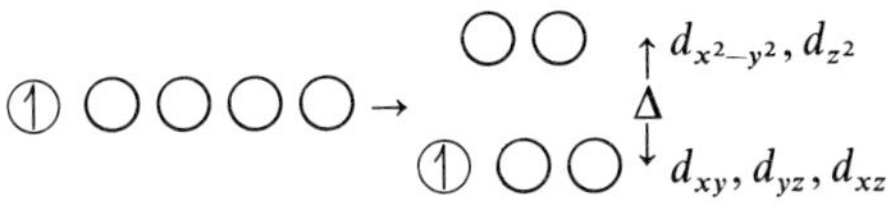

```
                      ○ ○   ↑ d_{x²-y²}, d_{z²}
① ○ ○ ○ ○  →                Δ
                  ① ○ ○   ↓ d_{xy}, d_{yz}, d_{xz}
```

The energy difference between the three lower-energy orbitals and the two upper-energy orbitals is called the *crystal-field splitting* and is given the symbol Δ.

A similar splitting of the d orbitals occurs when the ligands are neutral molecules as in the Ti $(OH_2)_6^{3+}$ ion. Here the electrostatic repulsions which give rise to the splitting are between the lone pairs of electrons on the oxygen atoms of water (p. 41) and the $3d$ electron on the Ti^{3+} ion.

Colours of transition-metal complexes

When light passes through a solution containing the $Ti(OH_2)_6^{3+}$ ion, a quantum of light may be captured by the ion, thus exciting an electron from a lower-level to a higher-level d orbital:

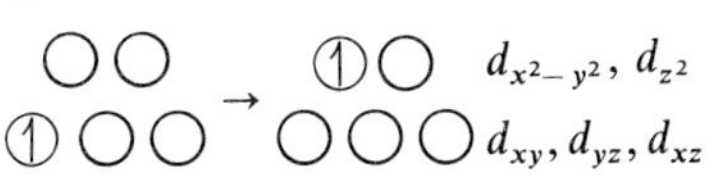

```
  ○ ○           ① ○    d_{x²-y²}, d_{z²}
          →
① ○ ○         ○ ○ ○  d_{xy}, d_{yz}, d_{xz}
```

The frequency ν of the light absorbed will be given by the expression

$$E = h\nu = \Delta,$$

the crystal-field splitting.

For the $Ti(OH_2)_6^{3+}$ ion, the crystal-field splitting is such that the light absorbed is in the visible part of the spectrum and the ion is coloured. In fact yellow light is absorbed and most of the blue and red light is transmitted; the result is that solutions containing the $Ti(OH_2)_6^{3+}$ ion are violet in colour.

It happens that the complexes of the first transition series usually absorb visible light and a great variety of colours are observed, the precise colour depending on the frequency of the light absorbed. Thus in aqueous solution, salts of chromium(III) are violet, those of manganese(II) are pale pink and those of copper(II) are bright blue.

In contrast, complexes of the elements of the second and third transition series are less highly coloured and are commonly pale yellow or white. For these elements the *d* orbitals project further into space towards the ligands so that the electrostatic repulsions between the ligands and the *d* electrons are increased. In turn this leads to a high crystal-field splitting, so that high-energy light (usually in the ultra-violet region of the spectrum) is absorbed, and the complexes are yellow or colourless.

Magnetic properties of transition-metal complexes

We have seen that transition-metal ions often have odd or unpaired *d* electrons; these impart a magnetic moment to the transition-metal ion. When a magnetic field is applied, this magnetic moment causes the transition-metal complex to be attracted into the stronger part of the field; the complex is said to be *paramagnetic*. Now the magnitude of this attractive force increases with the number of unpaired electrons on the transition-metal ion. Thus complexes of the Ti^{3+} ion having a single unpaired *d* electron interact less strongly with a magnetic field than complexes of V^{3+} which have two unpaired *d* electrons.

Let us consider the electronic configurations of transition-metal ions having several *d* electrons. For example, the Fe^{3+} ion has five 3*d* electrons. In octahedral complexes of this ion, two probable electron configurations are A and B:

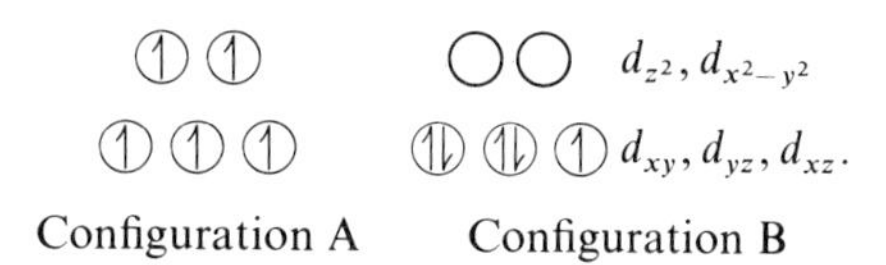

As noted earlier (p. 11) the electrons prefer to keep their spins parallel, and a definite energy, *P*, is required to pair the electron spins. Therefore if the crystal-field splitting Δ is small, the electron configuration A with five unpaired electrons will be adopted, while if Δ is greater than the pairing energy *P*, the electron configuration B with only one unpaired electron will be preferred.

Now in octahedral iron(III) complexes the *high-spin* electron configuration A is almost always adopted and a corresponding high paramagnetism is found. However, since the crystal-field splitting Δ is much greater for complexes of the second and third transition-series elements the *low-spin* electron configuration B is always found in octahedral complexes of ruthenium(III) and osmium(III). These complexes therefore have a low magnetic moment and are more weakly paramagnetic than the iron(III) complexes. The presence of more unpaired electrons in complexes of the first transition-series elements compared with those of the second and third transition series is a quite general phenomenon when alternative high-spin or low-spin electron configurations are possible.

Summary of transition-metal properties

We have seen that the transition elements have the common property of forming compounds in several oxidation states, but that the elements of the first transition series differ in several respects from the heavier transition elements. The differences which we have discussed in this chapter may be listed as follows:

(1) The first transition-series elements form high oxidation-state complexes less readily than the heavier transition metals. In particular the oxidation state two is much more common in the first transition series than in subsequent series.

(2) The complexes of the first transition-series elements exhibit a much larger range of colours and are often more strongly paramagnetic than those of the heavier elements.

(3) Covalent bonding is less important for the first transition-series elements in low oxidation states than for the second and third series. The heavier elements are more likely to exhibit 'Class *b*' metal behaviour, and so form water-insoluble halides and sulphides, than the elements of the first transition series.

PROBLEMS

7.1. Why are scandium and zinc not usually treated as transition metals?

7.2. Explain why 4*s* rather than 3*d* electrons are preferentially lost on forming transition metal *cations*, while the 4*s* are filled before the 3*d* orbitals during the building up of the periodic table to form the neutral atoms.

7.3. The ion Co^{3+} has the electron configuration (Ar core)$3d^6$ and has four unpaired electrons, yet there are no unpaired electrons in the complex ion $Co(NH_3)_6^{3+}$. How can you explain this? (Hint: see p. 72.)

7.4. Which of the following pairs of compounds do you expect to be the more soluble in water? Why?

(a) LiF, LiI (b) AgF, AgI

8. Periodicity and Chemical Compounds: The Lanthanides and Actinides

THE lanthanides and actinides are the elements in the periodic table which are formed as the $4f$ or $5f$ orbitals are progressively filled. There are therefore fourteen elements in each series, and all are metallic in nature.

The lanthanide elements and their ground-state electron configurations are listed in Table 13. The rather irregular trends in the electron configurations of these elements indicate that the valence orbitals (the $4f$, $5d$, and $6s$ orbitals) are very close to each other in energy. Table 13 also provides evidence for the increased stability of electron configurations having half-filled ($4f^7$) or filled ($4f^{14}$) shells of f electrons.

On forming positively charged ions, the $4f$ orbitals are strongly stabilized with respect to the other valence orbitals ($5d$ and $6s$). In fact this stabilization of the $4f$ orbitals in the positively charged lanthanide elements is so great that, by the time the tripositive lanthanide ions have been formed, the $4f$ electrons have entered the electron core, with the result that the chemistry of the lanthanide elements is dominated by the formation of M^{3+} ions. In the actinide elements the energies of the $5f$ orbitals are less sensitive to increases in nuclear charge on forming positively charged ions, so that oxidation states

TABLE 13

Ground-state electron configurations of the lanthanide elements

Element	Symbol	Ground state		
		$4f$	$5d$	$6s$
Lanthanum	La	0	1	2
Cerium	Ce	1	1	2
Praseodymium	Pr	3		2
Neodymium	Nd	4		2
Promethium	Pm	5		2
Samarium	Sm	6		2
Europium	Eu	7		2
Gadolinium	Gd	7	1	2
Terbium	Tb	9		2
Dysprosium	Dy	10		2
Holmium	Ho	11		2
Erbium	Er	12		2
Thulium	Tm	13		2
Ytterbium	Yb	14		2
Lutetium	Lu	14	1	2

higher than three can be reached more readily by these elements than by the lanthanides.

After this brief introduction, the chemistry of the lanthanide elements will be discussed and this will then be compared with the rather different chemistry of the actinide elements.

Binding energies of the lanthanide elements

The binding energies and melting points for the lanthanide elements are given in Fig. 29. It can be seen that the binding energies are almost constant on crossing the series; the magnitudes indicate that for all the elements there are just three electrons which contribute to the binding. Bearing in mind the properties of the transition elements (p. 60), it is not surprising that each lanthanide element also uses just three valence electrons in chemical combination with other elements.

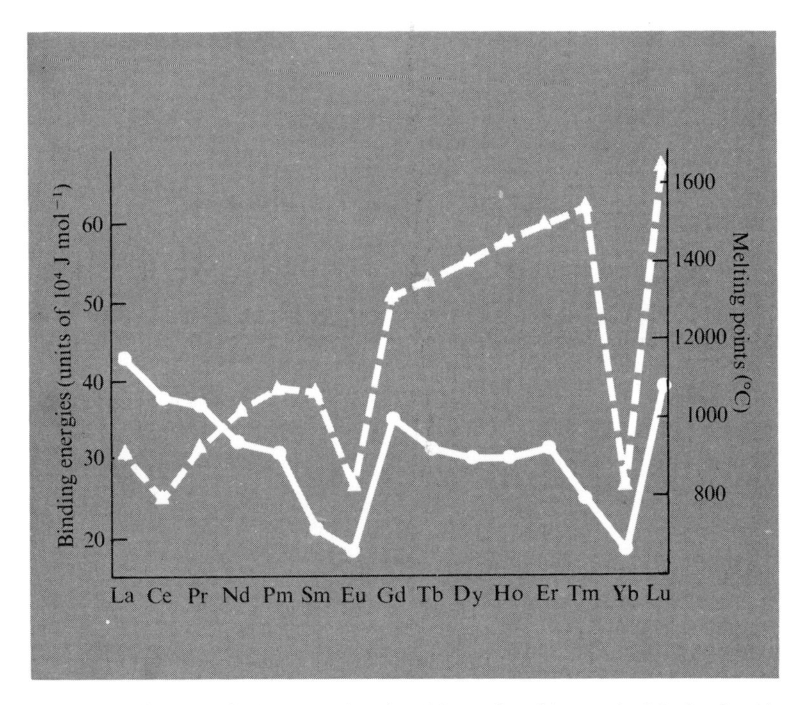

FIG. 29. Binding energies (○———) and melting points (△----) of the lanthanide elements.

Oxidation states of the lanthanides

The most stable oxidation state for all the lanthanide elements is three. Thus they all form ionic fluorides of general formula MF_3 and basic oxides M_2O_3.

For most of the lanthanides the *only* oxidation state is three, but there are certain exceptions. Europium and ytterbium can also form dipositive ions, Eu^{2+} and Yb^{2+}, which have the respective electron configurations (Xe core) $4f^7$ and (Xe core) $4f^{14}$ with favourable half-filled or filled shells of $4f$ electrons. Similarly cerium and terbium can form compounds in oxidation state four; the respective electron configurations for Ce^{4+} and Tb^{4+} are (Xe core) $4f^0$ and (Xe core) $4f^7$ respectively. Even in these cases however, the oxidation state three is the more stable. Thus compounds of Eu(II) are easily oxidized to Eu(III), while Ce(IV) compounds are easily reduced to Ce(III).

Covalent and ionic radii of the lanthanides

The covalent radii for the lanthanide elements and the radii of their tripositive ions are shown in Fig. 30. In each case there is a general decrease in

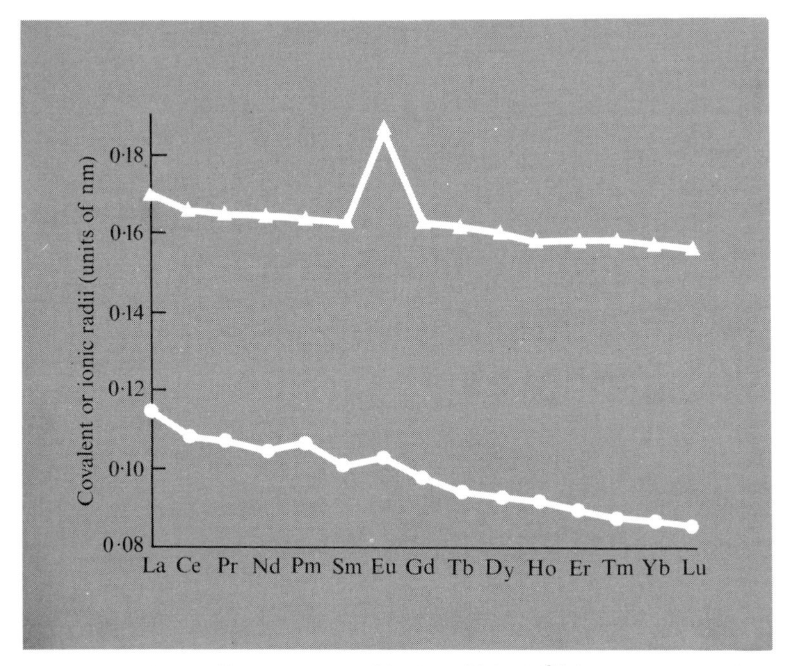

FIG. 30. Covalent radii (△———) and ionic radii for M^{3+} ions (○———) of the lanthanides.

the radii on passing across the lanthanide series due to the general increase in effective nuclear charge as the $4f$ electron shell is filled. This arises because the $4f$ electrons shield each other from the nuclear charge rather poorly, so that as more $4f$ electrons are added the effective nuclear charge felt by each of them increases. The result is that the whole $4f$ electron shell contracts on passing across the lanthanide series. This steady decrease in covalent and ionic radii on crossing the series is generally called the *lanthanide contraction.*

We have already seen (p. 65) that the lanthanide contraction is responsible for the similarity in covalent and ionic radii between the elements of the second and third transition series. The effect is also important in the chemistry of the lanthanides themselves, since the general decrease in ionic radii on crossing the series leads to a corresponding increase in the polarizing power of the ions and in the stability of complexes of the ions.

The actinide elements

The actinide elements are all radioactive and this has prevented a thorough investigation of the properties of many of them. The elements and their probable ground-state electron configurations are given in Table 14.

As might be expected, the actinides have much in common with the lanthanides, but there are also important differences. These arise because the $5f$ orbitals of the actinide elements do not drop in energy to such a great

TABLE 14

Ground-state electron configurations of the actinide elements

Element	Symbol		Ground state		
			$5f$	$6d$	$7s$
Actinium	Ac			1	2
Thorium	Th			2	2
Protoactinium	Pa		2	1	2
		or	1	2	2
Uranium	U		3	1	2
Neptunium	Np		5		2
Plutonium	Pu		6		2
Americium	Am		7		2
Curium	Cm		7	1	2
Berkelium	Bk		8	1	2
		or	9		2
Californium	Cf		10		2
Einsteinium	Es		11		2
Fermium	Fm		12		2
Mendelevium	Md		13		2
Nobelium	No		14		2
Lawrencium	Lw		14	1	2

TABLE 15

Oxidation states of the actinide elements. The most important oxidation state(s) for each element are underlined

Ac	Th	Pa	U	Np	Pu	Am	Cm	Bk	Cf	Es	Fm	Md	No
3	3	3	3	3	3	3	3	3	3	3	3	3	3
	4		4	4	4	4	4	4					
		5	5	5	5	5							
			6	6	6	6							

extent as do the $4f$ orbitals of the lanthanides when positive ions are formed. For this reason, the actinide elements often form compounds in which their oxidation states are higher than three.

Oxidation states of the actinide elements

The important oxidation states of the actinide elements are shown in Table 15. It can be seen that, like the lanthanides, the actinides all form compounds in oxidation state three, but that higher oxidation states are possible and indeed are favoured at the start of the series. Thus thorium(IV), protoactinium(V), and uranium(VI) compounds are all more stable than the compounds of these elements in oxidation state three. At the start of the actinide series then, the behaviour is more reminiscent of the transition elements than of the lanthanides, but later in the series the oxidation state three becomes progressively more important until at the end of the actinide series the properties of the elements and their compounds are very similar to those of the lanthanides.

Other similarities with the lanthanides can be attributed more directly to the presence of partially-filled shells of f electrons. Thus, like the transition-element complexes, the compounds of both the lanthanides and actinides are often coloured and paramagnetic.

PROBLEMS

8.1. The element with atomic number 104 has recently been detected. Can you predict any of its chemical properties from its expected position in the periodic table?

Bibliography and Notes

Valency and Molecular Structure, 3rd ed., by E. CARTMELL and G. W. A. FOWLES. Butterworths, London (1966).
This book gives a thorough treatment of atomic structure and the various applications of quantum theory to the problems of chemical bonding. The stereochemistry and bonding in the compounds of the elements are then discussed.

Introduction to Modern Inorganic Chemistry, by K. M. MACKAY and R. A. MACKAY. Intertext Books, London (1968).
A good coverage of all aspects of inorganic chemistry at a level suitable for first and second year university students is given in this text.

Advanced Inorganic Chemistry, 2nd ed., by F. A. COTTON and G. WILKINSON. Interscience Publishers (1966).
This is the best single textbook on inorganic chemistry, giving an excellent account of the principles of chemical combination with a more detailed treatment of the chemistry of individual elements.

Inorganic Chemistry, by C. S. G. PHILLIPS and R. J. P. WILLIAMS. Clarendon Press, Oxford (1966).
This is an excellent advanced text on inorganic chemistry in two volumes. It is especially good in emphasizing periodic trends in chemical behaviour. The treatment is very rigorous but the book can be recommended for the above-average undergraduate. Many of the ideas presented in a simplified form in the present text are developed at length by Phillips and Williams.

Co-ordination Chemistry, by F. BASOLO and R. C. JOHNSON. W. A. Benjamin, Inc., New York, Amsterdam (1964).
This very readable book deal only with transition-metal complexes but is nevertheless very useful.

Principles of Organometallic Chemistry, by G. E. COATES, M. L. H. GREEN, P. POWELL, and K. WADE. Methuen and Co. Ltd., London (1968).
Compounds containing metal–carbon bonds of many different types are discussed in this stimulating paperback text.

Problems in Inorganic Chemistry, by B. J. AYLETT and B. C. SMITH. The English Universities Press Ltd., London (1965).
Test how much you have learned from the above texts by trying these often numerical questions. Answers are given.

Index

Answers to problems

2.1. (a) $n = 2, l = 0, m = 0,$ $s = +\frac{1}{2}$ or $-\frac{1}{2}$
(b) $n = 2, l = 1, m = 1, 0,$ or $-1,$ $s = +\frac{1}{2}$ or $-\frac{1}{2}$
(c) $n = 3, l = 2, m = 2, 1, 0, -1,$ or $-2,$ $s = +\frac{1}{2}$ or $-\frac{1}{2}$

Two electrons can be accommodated in the 2*s* orbital, six electrons in the 2*p* orbitals, and ten electrons in the 3*d* orbitals.

3.1.

Atomic number	Electron configuration	Element	Group	Block
4	$1s^2 2s^2$	Be	II	*s*-block
16	$1s^2 2s^2 2p^6 3s^2 3p^4$	S	VI	*p*-block
22	$1s^2 2s^2 2p^6 3s^2 3p^6 3d^2 4s^2$	Ti	IVA	*d*-block

4.1. Ionization energy, Electronegativity: (a) $B < N < F < Ne$; (b) $Ba < Sr < Ca < Mg < Hg$

4.2. $Na^+ > Li^+ > Be^{2+} > B^{3+}$. B^{3+} is formed least readily.

4.3.

Compound	Oxidation state	Co-ordination number
$SiCl_4$	4	4
$SiCl_6^{2-}$	4	6
$AlCl_4^-$	3	4
AlF_6^{3-}	3	6

5.2. (a) $Si > Al > Mg > Na > K$. This is the order of decreasing binding energies.
(b) $S > P > Cl > Ar$. This is the order of decreasing molecular size; i.e. $S_8 > P_4 > Cl_2 > Ar$.

5.3. (a) $Al > Si > P$
(b) $Po > Te > Se$

6.2. BCl_3, $SiCl_4$, and PCl_5 are hydrolysed rapidly by water

7.3. In $Co(NH_3)_6^{3+}$, cobalt has the electron configuration ( ) ( ) above (↑↓) (↑↓) (↑↓) with no unpaired electrons, while the free ion Co^{3+} has the configuration (↑↓) (↑) (↑) (↑) (↑), with four unpaired electrons.

7.4. Solubility of $LiI > LiF$ but $AgF > AgI$. Lithium is a 'class a' metal, while silver is 'class b'.

8.1. Element 104 falls into group IVA and should resemble Ti, Zr, and Hf in its chemical properties.